AF869235

Springer Series in Reliability Engineering

Series Editor

Hoang Pham, Industrial and Systems Engineering, Rutgers, The State University of New Jersey, Piscataway, NJ, USA

Today's modern systems have become increasingly complex to design and build, while the demand for reliability and cost effective development continues. Reliability is one of the most important attributes in all these systems, including aerospace applications, real-time control, medical applications, defense systems, human decision-making, and home-security products. Growing international competition has increased the need for all designers, managers, practitioners, scientists and engineers to ensure a level of reliability of their product before release at the lowest cost. The interest in reliability has been growing in recent years and this trend will continue during the next decade and beyond.

The Springer Series in Reliability Engineering publishes books, monographs and edited volumes in important areas of current theoretical research development in reliability and in areas that attempt to bridge the gap between theory and application in areas of interest to practitioners in industry, laboratories, business, and government.

Now with 100 volumes!

Indexed in Scopus and EI Compendex

Interested authors should contact the series editor, Hoang Pham, Department of Industrial and Systems Engineering, Rutgers University, Piscataway, NJ 08854, USA. Email: **hopham@soe.rutgers.edu, or Anthony Doyle, Executive Editor, Springer, London. Email: anthony.doyle@springer.com**.

Hongyan Dui · Shaomin Wu

Importance-Informed Reliability Engineering

Springer

Hongyan Dui
School of Management
Zhengzhou University
Zhengzhou, Henan, China

Shaomin Wu
Kent Business School
University of Kent
Canterbury, Kent, UK

ISSN 1614-7839 ISSN 2196-999X (electronic)
Springer Series in Reliability Engineering
ISBN 978-3-031-52454-7 ISBN 978-3-031-52455-4 (eBook)
https://doi.org/10.1007/978-3-031-52455-4

© The Editor(s) (if applicable) and The Author(s), under exclusive license to Springer Nature Switzerland AG 2024

This work is subject to copyright. All rights are solely and exclusively licensed by the Publisher, whether the whole or part of the material is concerned, specifically the rights of translation, reprinting, reuse of illustrations, recitation, broadcasting, reproduction on microfilms or in any other physical way, and transmission or information storage and retrieval, electronic adaptation, computer software, or by similar or dissimilar methodology now known or hereafter developed.
The use of general descriptive names, registered names, trademarks, service marks, etc. in this publication does not imply, even in the absence of a specific statement, that such names are exempt from the relevant protective laws and regulations and therefore free for general use.
The publisher, the authors, and the editors are safe to assume that the advice and information in this book are believed to be true and accurate at the date of publication. Neither the publisher nor the authors or the editors give a warranty, expressed or implied, with respect to the material contained herein or for any errors or omissions that may have been made. The publisher remains neutral with regard to jurisdictional claims in published maps and institutional affiliations.

This Springer imprint is published by the registered company Springer Nature Switzerland AG
The registered company address is: Gewerbestrasse 11, 6330 Cham, Switzerland

Paper in this product is recyclable.

This book is dedicated to
my wife Yachen, my child Wenjin, and my parents

H. Dui

my wife Huiqing, my children Siqi and Eddy, and my parents

S. Wu

Preface

Reliability engineering is a discipline for applying operational research, statistics, and computer sciences to ensure an engineered system to perform its intended functions for the required time duration in a specified environment. Due to the technological advance, engineered systems are becoming more complex and maintenance on them is becoming more challenging. Different methods have therefore developed to keep a system at a certain level of availability and to facilitate the maintenance. Reliability importance measures are one of those methods.

Reliability importance measures are proposed to prioritise system components from different perspectives. The lifecycle of an engineered system can be divided into different stages, including the design stage, the production stage, the operation stage and the disposal stage. Reliability importance measures can be used at different stages for improving a system's reliability, availability, and save the operating costs, to name a few. Reliability importance measures have shown their wide spectrum of applications in the lifetimes of engineered systems.

There are enormous publications relating to importance measures in reliability mathematics and engineering related journals and conferences. Especially, in the last 10 years, more research has been done in this area and more work has been published. Hence, there is a need to publish a book for updating the latest development relating to reliability importance measures and their applications in engineering, which is the incentive and motivation of writing this book.

The book is developed based on the authors' research work, on which the first and second authors have published more than fifty papers and a dozen of papers, respectively. Their work is mainly on extensions of the Birnbaum importance measure and their applications.

From the content perspective, this book can be divided into three parts. Part 1, which only contains Chapter 1, introduces some concepts that are needed to the other chapters. Part 2, including the five chapters (i.e., Chaps. 2–6), introduces importance measures and investigates their various properties. Part 3, which only includes Chap. 7, provides some cases for illustrating the importance measures discussed in Part 2. In more detail about each chapter in Part 2, Chap. 2 discusses importance measures for system design. Chapters 3 and 4 investigate the application of importance measures on maintenance. Chapter 5 extends the conventional definitions of importance measures to the scenarios of networks, which are composed of edges and nodes. Chapter 6 applies importance measures to resilience management.

The prerequisite knowledge for reading this book is an undergraduate calculus course and a probability and statistics course. The book is also a good reference for graduate students and researchers in the areas of operational research and engineering such as mechanic engineering, transport engineering and civil engineering, to name a few.

Zhengzhou, China
Canterbury, UK
24 October, 2023

Hongyan Dui
Shaomin Wu

Acknowledgements

We are grateful to Mr. Xinghui Dong, Miss Songru Zhang, Miss Yulu Zhang, Miss Jiaying Song, Miss Xinmin Wu, Miss Xinyue Wang, Miss Kaixin Liu, Mr. Yaohui Lu, Mr. Jiabao Zhai, Mr. Jiafeng Wang, Miss Weina Guo, Miss Hao Zhang, Miss Huiting Xu, Mr. Xuxing Wei, and Mr. Xiao Wang for their assistance in preparation of the manuscript. We should also express our thanks to the colleagues of Springer Nature for their effort and work so that this book can be prosperously published.

The first author is appreciative of the research fund granted by the National Natural Science Foundation of China (No. 72071182).

Contents

Acronyms

ABAO	As Bad As Old
AGAN	AGAN
ARP	Age Replacement Policy
BIM	Birnbaum Importance Measure
BRP	Block Replacement Policy
CCMI	Cost-based Component Maintenance Importance
CIP	Cost-based Improvement Potential
CM	Corrective Maintenance
CMP	Component Maintenance Priority
DIM	Differential Importance Measure
GP	Geometric Process
IIM	Integrated Importance Measure
IPI	Increased Potential Importance
JDIM	Joint Differential Importance Measure
JFI	Joint Failure Importance
JIIM	Joint Integrated Importance Measure
LIC	Level of Irregularity of a Cluster
LIN	Level of Irregularity of a Node
MAD	Mean Absolute Deviation
MCIM	Multi-Criteria Importance Measure
MEM	Maintenance Efficiency Measure
MER	Matrix of Edge Resilience
MNR	Matrix of Node Resilience
NC	Node Capacity
NHPP	Non-Homogeneous Poisson Process
OM	Operations and Maintenance
PM	Preventive Maintenance
RBD	Reliability Block Diagram
RCF	Relative Capacity of the Flow
RFC	Relative Flow Capacity

RE	Resilience of Edge
REIM	Resilience Efficiency Importance Measure
REIMII	An extension of REIM to evaluate the rehabilitation benefits of components under different maintenance sequences
RIM	Recovery Importance Measure
RLC	Relative Level of Circulation
RLT	Ratio of Real-Time Load Capacity
RP	Renewal Process
RTR	Derivative of RLT
RVATP	Rocket Vertical Assembly And Test Plant
VIM	Vulnerability Importance Measure

Annotation

$(\cdot_{i,X(t)})$	$(X_1(t), \ldots, X_{i-1}(t), \cdot, X_{i+1}(t), \ldots, X_n(t))$
$(\cdot_i, \boldsymbol{p})$	$(p_1, \ldots, p_{i-1}, \cdot_i, p_{i+1}, \ldots, p_n)$
$(\cdot_i, \cdot_j, \boldsymbol{p})$	$(p_1, \ldots, p_{i-1}, \cdot_i, p_{i+1}, \ldots, p_{j-1}, \cdot_j, p_{j+1}, \ldots, p_n)$
$\lambda_i(t)$	Failure rate of component i
∇f_g	Component attribute gradient vector under the f_g criterion
$\phi(x_k)$	Attribute characteristic vector for x_k
$\Phi(\boldsymbol{X})$	System structure function
$\Phi(\boldsymbol{X}(t))$	$\Phi(\boldsymbol{X}(t)) = \Phi(X_1(t), X_2(t), \ldots, X_n(t))$
$\phi(\tau, \boldsymbol{p})$	System reliability corresponding to the permutation τ
ρ_{i_m}	$\Pr\{X_i \geq m\} = P_{i_m} + P_{i_{(m+1)}} + \cdots + P_{i_{M_i}}$
τ	Mutation $(\tau(1), \tau, \ldots, \tau(n))$ of the integers from 1 to n
τ_{ij}	Permutation obtained from a permutation τ by interchanging components in positions $\tau(i)$ and $\tau(j)$
α_{ij}	Correlation between two criteria
a_j	Performance level corresponding to state j of the system
A	The adjacency matrix
B	Ideal characteristic vector
c_j	Maintenance cost corresponding to state j of the system
d_{ij}	Shortest path between node i and node j in the network
E	Set of edges
f_g	g^{th}criterion in a multi-criterion engineered system
G	Mono-layer network composed of nodes and edges
$k_{i,\text{out}}$	Number of out-degrees of node i
$k_{i,\text{in}}$	Number of in-degrees of node i
k_{in}^{Cl}	Number of in-degrees of a cluster
k_{out}^{Cl}	Number of out-degrees of a cluster
L_i	Initial capacity of node i
V	Set of vertices
I_i^B	Birnbaum importance of component i in binary systems
$I_{i_m}^G$	Griffith importance of state m of component i

I_i^{im}	IIM value of component i in binary systems
I^{mc}	Multi-Criteria importance measure
$I_m^e(t)^{S_{N'}}$	Resilience efficiency importance of component m
I_i^{IP}	Increase potential importance of component i
I_i^j	Joint importance of component i
I_i^{jim}	Joint integrated importance of component i
$I_{i,j}^j$	Joint importance of component group (i, ..., j) in consecutive k-out-of-n: F system
I^D	Differential importance
I^v	Vulnerability importance
I^m	Maintenance efficiency measure
I^M	Component maintenance priority
I^r	Recovery importance measure
I_i^{mad}	Mean absolute deviation of component i
I^c	Cost-based importance
I^{cim}	Cost-based IIM
I^C	Cost-based component maintenance importance
I^P	Cost-based improvement potential importance measure
I^{pm}	Cost-based PM policy
I^f	Joint failure importance
I^{om}	Cost-based OM policy
J_{gk}	Sensitivity of the x_k's attribute to the system attribute under f_g criterion
J'_{gk}	Standardized J_{gk}
k	Minimum number of consecutive failing components, which cause system failure
M	Number of system states
M_i	Number of states of component i with $i = 1, 2, \ldots, n$
n	Number of components in a system
N'	Set of failed components
$N_{j-i+1,k}^{(i,j)}$	Number of no overlapping failure runs of length k among the components$i$, ..., j
N_n, k	Total number of no overlapping runs of failures of length k
N^{lic}	The Level of irregularity of a cluster
N_i^{lin}	The level of irregularity of a node i
N_{ij}^{rlc}	Static property that measures the continuity of node i and node j
N_{ij}^{rlt}	Relative load transfer rate between node i and node j
N_{ij}^{rcf}	Relative flow capacity of an edge between node i and node j
N_i^{nc}	Capacity of a node i
N_i^{nms}	Maintenance policy of node i
$N^{ems_f^e}$	The maintenance policy of edge f when any edge e fails
N^{otms}	Overall maintenance strategy of the network
N^{onms}	Maintenance strategy of nodes

N^{tems}	Maintenance strategy of edges
N_i^{rq}	Relative real time load of node i
N_i^{qrn}	Relative rate of real-time load transfer of node i
N^{MNR}	Node resilience matrix
N^{MER}	Resilience matrix of Edges
$N^{re_i^j}$	Rate of change of the value of edge i when edge j fails
p_i	Probability that component i functions, $i = 1, 2, \ldots, n$
$p_{\tau(i)}$	Reliability of component residing in particular position $\tau(i)$
$\boldsymbol{p}$	$(p_1, p_2, \ldots, p_n)$, component reliability vector
$P_{i1}(t)$	Reliability of component i, $P_{i1}(t) = \Pr\{X_i(t) = 1\}$
P_{im}	$\Pr\{X_i = m\}, m = 0, 1, 2, \ldots, M_i\}$
$P_i^{CIM}(C)$	The average maintenance cost of component i in a maintenance cycle
$P_i^{ECL}(T)$	The expected maintenance cycle length of component i
$P_i^{EMT}(T)$	The expected maintenance time of component i
$P_i^{EOT}(T)$	The expected operational time of component i
q_i	$p_i + q_i = 1$
Q_{im}	$Q_{im} = \Pr\{x_i \leq m\} = P_{i0} + \cdots + P_{im}$
q_i	$1 - p_i$
$R(\boldsymbol{p})$	System reliability for component reliability vector $\boldsymbol{p}$
$R(n; k)$	Reliability of a linear consecutive-k-out-of-n system Furthermore, $R(n; k) + Q(n; k) = 1$
$R(p, k, n)$	Reliability of k-out-of-n: F system
$R(p, k, n - j)$	Reliability of k-out-of-n: F subsystem consisting of components $(n - j+1), \ldots, n$
$R^t(j)$	Reliability of consecutive-k-out-of-j: F subsystem consisting of components 1, 2,...,j
$R_i(t)$	Availability of component i at time t,$R_i(t) = Pr\{X_i(t) = 1\}$
$R_c(p, k, n)$	Reliability of circular$k$-out-of-n: F system
$RL(p, k, n)$	Reliability of linear k-out-of-n: F system
S	Set of all possible permutations
$S_{j-i+1,k}^{(i,j)}$	Number of working components among the components i, ..., j
$S_{N'}$	Set of the failure states of failed components in N'
U	Expected performance of a system
$U(\mathbf{X}(t))$	Expected performance of a system at time t
$v_g(\boldsymbol{X})$	Attribute of a system under the $f_g$criterion
W	Event that system works
w_g	Weight of criterion f_g
$\boldsymbol{X}$	$\boldsymbol{X} = (X_1, X_2, \ldots, X_n)$: state vector of the n components
$\boldsymbol{X}(t)$	$(X_1(t), X_2(t), \ldots, X_n(t))$: state vector of the components
$X_i(t)$	State of component i at time t, $X_i(t) = 0, 1, 2, \ldots, M_i$
X_i	State of component i,$X_i = 0, 1, 2, \ldots, M_i$
x_k	kth component in a multi-criterion engineered system.

Chapter 1
Introduction

Abstract Reliability mathematics aims to use a multidisciplinary approach to improving the reliability of engineered systems and avoiding negative consequences due to system failures. In this subject, operational research, probability, and statistics are widely used approaches in tackling various challenges. This chapter introduces basic concepts that are commonly used in reliability engineering.

Keywords Reliability · Failure rate function · k-out-of-n system · Maintenance effectiveness · Maintenance policy · Importance measures · Resilience

1.1 Basic Concepts of Reliability

1.1.1 Reliability Function

In many real applications, time to the occurrence of an event is a main concern. For example, time to occurrence can be regarded as time to failure for an engineered product, time to death for patients who have taken a newly developed medicine, time to buy your first home, among others. If we denote T as time to failure and $f(t)$ as its probability density function (pdf). Then $f(t)$ describes how the failure probability is spread over time.

Apparently,

$$\int_0^\infty f(t)dt = 1,$$

which means the total area under the pdf curve (which is the curve of $f(t)$) must always equal 1 and suggests that the system or component will certainly fail in the long run.

Let $F(t)$ denote the cumulative distribution function (cdf) of T. Then $F(t)$ is the probability $\Pr(T \leq t)$, which is the probability that the system will fail by time t:

© The Author(s), under exclusive license to Springer Nature Switzerland AG 2024

H. Dui and S. Wu, *Importance-Informed Reliability Engineering*, Springer Series in Reliability Engineering, https://doi.org/10.1007/978-3-031-52455-4_1

$$F(t) = \Pr(T \le t) = \int_0^t f(v)dv.$$

The relationship between pdf and cdf is

$$f(t) = \frac{dF(t)}{dt}.$$

The probability $\Pr(t_1 < T \le t_2)$, which is the probability that the system will fail between times t_1 and t_2, is

$$\Pr(t_1 < T \le t_2) = \int_{t_1}^{t_2} f(t)dt = F(t_2) - F(t_1).$$

Definition 1.1 *(Reliability)* Reliability is the ability of an entity to perform a required function under given conditions for a given time interval [5].

It should be stressed that the key words in the above definition include *required function*, *given condition*, and *given time interval*. Accordingly, failure is the termination of the ability to perform the required function.

If we interpret the above definition using the language of probability, reliability is the probability $\Pr(T > t)$, namely, reliability is the probability that the time to failure T is greater than a specified time t. If we denote $R(t)$ as the reliability function, then

$$R(t) = \Pr(T > t) = 1 - F(t).$$

$R(t)$ is also referred to as the survival function. That is, reliability is the probability that an item works without failure during a time interval $(0, t)$ under given usage intensity and operating environment. An *item* or an *entity* can be a component, a subsystem, or a system.

From Definition 1.1, it can be seen that

- $R(t_1) \ge R(t_2)$ if $t_1 < t_2$. That is, $R(t)$ is a non-increasing function of t. With time progressing, the reliability of an item becomes smaller.
- $R(0) = 1$. That is, at time $t = 0$, the reliability of a new item is 1.
- $R(\infty) = 0$. That is, when time approaches infinite, an item will definitely fail.

1.1.2 Failure Rate Function

Another importance concept in reliability mathematics is the hazard rate, or hazard function, which is defined as follows.

Definition 1.2 *(Failure rate (or hazard function))* Given that an item has survived time t, the probability of failure within the small time interval $(t, t + \Delta t)$ is denoted by $h(t)\Delta t$. Then $h(t)$ is the hazard rate or hazard function.

The relationship between the reliability and the failure rate is given below:

$$f(t)\Delta t = R(t)h(t)\Delta t,$$

from which an expression for the failure rate $h(t)$ can be obtained:

$$h(t) = \frac{f(t)}{R(t)}.$$

Since $R(t) \leq 1$, the failure rate $h(t)$ will always be greater than $f(t)$. Furthermore, $f(t) = -R'(t)$, $\frac{R'(t)}{R(t)} = -h(t)$, then integrating both sides with the initial condition $R(0) = 1$ gives an expression which relates the reliability with the failure rate:

$$R(t) = \exp(-\int_0^t h(v)dv).$$

The reliability $R(t)$ of an item can be increased by decreasing the failure rate $h(t)$, which decreases the value of the integral $\int_0^t h(v)dv$.

1.1.3 Reliability Bath-Tub Curve

In the lifetime of an item, the failure rate $h(t)$ of the item may experience three periods: *infant mortality failure period*, *random failure period*, and *wear-out failure period*, which constitutes a curve known as the bathtub curve, as illustrated in Fig. 1.1.

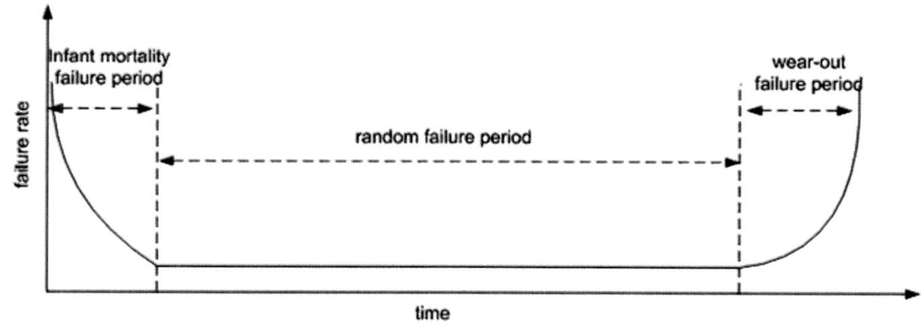

Fig. 1.1 Bathtub curve

The three periods in the bathtub curve can be interpreted as follows.

- The first period, which is also referred to as the *infant mortality failure period*, is characterised by a decreased failure rate.
- The second period, or the *random failure period*, is characterised by an approximately constant failure rate. Failures in this period are not due to age, wear-out or degradation. It is important to note that an intervention such as preventive maintenance does not affect the failure rate during this period.
- The third period, or the *wear-out failure period*, is characterised by an increased failure rate due to wear-out and degradation of properties.

Each of the three periods of the bathtub curve can be described by the Weibull distribution

$$F(t) = 1 - \exp(-(t/\eta)^{\beta}).$$

where η is the scale parameter and β is a shape parameter.

If $\beta = 1$, the Weibull distribution reduces to the negative exponential distribution

$$F(t) = 1 - \exp(-t/\eta).$$

It is easy to obtain the probability density function of the Weibull distribution:

$$f(t) = \frac{\beta}{\eta}(\frac{t}{\eta})^{\beta-1}\exp[-(t/\eta)^{\beta}].$$

Since the failure rate is given by $h(t) = f(t)/R(t)$, the Weibull failure rate function becomes

$$h(t) = (\beta/\eta)(\beta/\eta)^{\beta-1}. \tag{1.1}$$

As can be verified from this equation,

- If $\beta < 1$, the failure rate is decreasing. The corresponding Weibull distribution can model the probability of failures of an item in the infant mortality failure period;
- If $\beta = 1$ (which corresponds to a constant failure rate), the corresponding Weibull distribution can model the probability of failures of an item in the constant failure period;
- for $\beta > 1$, the failure rate is increasing. The corresponding Weibull distribution can model the probability of failures of an item in the wear-out period.

1.2 System Reliability Analysis

Assume a system is composed of multiple components and the reliability of each component is known. Then we will be able to obtain the reliability of the system, given the assumption that the lifetimes of the components are statistically independent.

1.2.1 Reliability of a Series System

Definition 1.3 *(Series system)* Suppose a system is composed of multiple components. If the failure of any one of the components causes the system to fail, then the system is said a series system.

For example, a car has four tyres. If one of the tyres deflates, then the tyre system stops working and the deflated tyre needs repairing. As such, the tyre system is a series system.

We may also create a figure to represent a system composed of multiple components. For example, assuming a system is composed of three components, 1, 2, and 3, then we can use Fig. 1.2 to illustrate a series system composed of these three components. Figure 1.2 is also called the reliability block diagram (RBD) of the series system.

Assume a series system is composed of n components that are statistically independent, where *statistically independent* means that the failure of a component does not affect the operating of other components in the system. Denote the lifetime distribution of component i as $F_i(t)$ for $i = 1, 2, \ldots, n$. According to the definition: if one of the components in a series system fails, the system fails. Alternatively, we can express this statement to be: to ensure the system works, all the components in the system need working. Then the reliability of the system is given by

$$R_s(t) = \prod_{i=1}^{n} R_i(t).$$

1.2.2 Reliability of a Parallel System

Definition 1.4 *(Parallel system)* Suppose a system is composed of multiple components. Assuming the system fails only if all the components in the system failed, then the system said a parallel system.

For example, a human body has two kidneys. Only if both kidneys fail, the kidney system will fail. As such, the kidney system can be regarded as a parallel system.

Assuming a system is composed of three components, 1, 2, and 3, then we can use Fig. 1.3 to illustrate the parallel system composed of these three components.

Assume a parallel system is composed of n components and they are statistically independent. Denote the lifetime distribution of component i as $F_i(t)$ for

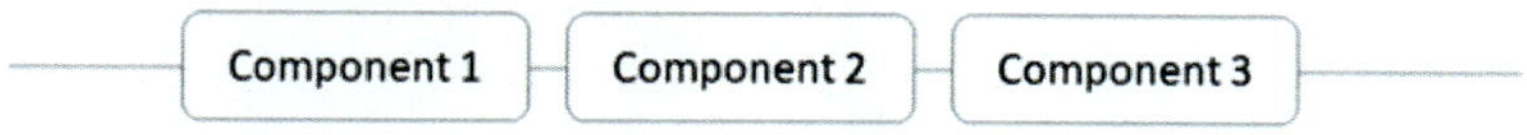

Fig. 1.2 A series system

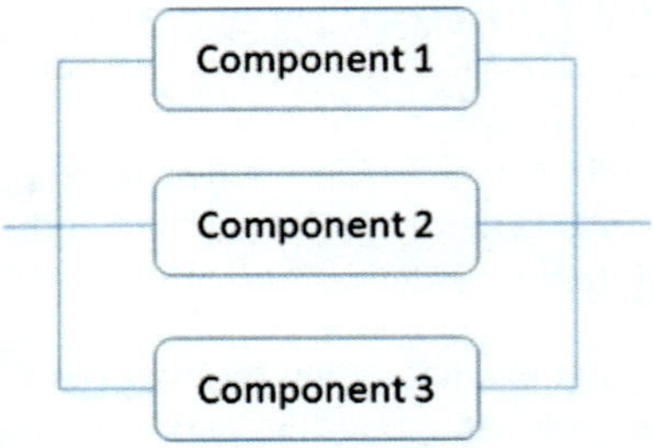

Fig. 1.3 A parallel system

$i = 1, 2, \ldots, n$. According to the definition: only if all the components in the parallel system fail, the system fails. Alternatively, we can express this statement to be: as long as one of the components in the system works, the system works. Then the lifetime distribution of the system is given by

$$F_s(t) = \prod_{i=1}^{n} F_i(t).$$

As such, the reliability of the system is given by

$$R_s(t) = 1 - \prod_{i=1}^{n} F_i(t) = 1 - \prod_{i=1}^{n} (1 - R_i(t)).$$

1.2.3 Reliability of a k-out-of-n System

Definition 1.5 *(k-out-of-nsystem)* Suppose a system is composed of n components. The system functions if and only if at least k components (out of the n components) functions successfully.

For example, a V8 engine is composed of 8 cylinders. The engine works if and only if more than four cylinders work. Then this engine is a 4 out of 8 system.

Assume that the components in a k-out-of-n system are statistically independent and that all components have the identical reliability function $R(t)$. Then the reliability function of the system is given by

$$R_S(t) = \sum_{i=k}^{n} \binom{n}{i} [R(t)]^i [1 - R(t)]^{n-i}.$$

1.2.4 *Reliability Improvement and Optimisation for Non-repairable Systems*

The lifetime of an item can normally be divided into several stages: design, manufacturing and installation, operation and maintenance, and disposal. At the design stage, the engineers may be required to meet the reliability goal for a product. They will then need to assign reliability to each individual component or subsystem to achieve this goal, considering constraints such as limited space, limited budget, and limited weight.

Once the product items are at the operation and maintenance stage, the engineers will need to improve the reliability by scheduling optimal replacement policies or preventive maintenance (PM) policies (see Sect. 1.3 for more discussion), which can be prioritized based on the rankings of importance measures, to improve the reliability of the system, both PM and importance measures are the focus of this book.

1.2.5 *Types of Engineered Systems*

In the reliability related literature, an engineered system can be one of the following types.

Binary system A binary system is always in one of the two states: it may be either working or failed. Light bulbs are a good example of this type.

Multi-state system A multi-state system can be working at a number of different states. For example, an offshore electrical power generation system can be regarded as a multi-state system [16] since it can generate different levels of power or can be regarded as working at a different state.

Continuum system A continuum system is a system whose performance is a continuous variable. An example of a continuum system is an automobile tire, the performance of which degrades [6].

It should be noted that it is logically meaningless to discuss the performance of a binary system as the system has either a 100% level of performance or a 0 level of performance.

1.3 Optimisation of Maintenance Policies for Items with Non-observable Failure Progression

When a system of interest is repairable and the failure progression is unobservable, we may use a stochastic point process to model the failure process of the system and then schedule preventive maintenance policies according to associated information such as the expected cost of failures and the expected cost of repair. Stochastic point processes are widely used to model times between failures. Such processes include the renewal process, the non-homogeneous Poisson process and the generalised renewal process. To use these models or processes, we need to understand the effectiveness of repair or maintenance.

1.3.1 Stochastic Processes for Modelling Times-Between-Failures

In reliability mathematics, the effectiveness of maintenance upon failure of an item is typically categorised into: perfect repair, imperfect repair, and minimal repair.

Perfect repair, in which a repair restores the condition of a failed item to an "as good as new (AGAN)" status. For example, a failed item is replaced with a new identical one. See Fig. 1.4 for an illustration. The renewal process is a widely used model for the failure process of items under perfect repair [21].

Definition 1.6 *(Renewal process)* [21] Given a sequence of random variables $\{X_k, k = 1, 2, \ldots\}$, if they are independent and the cdf of X_k is given by $F(x)$ for $k = 1, 2, \ldots$, then $\{X_k, k = 1, 2, \ldots\}$ is called a renewal process (RP).

Minimal repair, in which a repair restores a failed item to its state immediately prior to the failure. The operating state of an item after minimal repair is often called "as bad as old (ABAO)" in the literature. See Fig. 1.5 for a virtual illustration. The only model of minimal repair available in the literature is the non-homogeneous Poisson process (NHPP).

Definition 1.7 *(Non-homogeneous Poisson process)* [21] The counting process $\{N(t), t \geq 0\}$ is said to be a non-homogeneous Poisson process with intensity function $\lambda(t), t \geq 0$, if it satisfies

- $N(0) = 0$ almost surely;
- $\{N(t)\}$ has independent increments;

Fig. 1.4 Perfect repair

Fig. 1.5 Minimal repair

Fig. 1.6 Imperfect repair

- $P[N(t+h) - N(t) \geq 2] = o(h)$; and
- $P[N(t+h) - N(t) = 1] = \lambda(t) + o(h)$.

Imperfect repair, in which a repair restores a failed item to a status somewhere between "as good as new" and "as bad as old". See Fig. 1.6 for a virtual illustration. Many models, including the geometric process (GP) and its variants [12, 26, 28], the generalised renewal process models [7], have been developed for modelling the failure process of an item under imperfect repair.

Definition 1.8 [12] Given a sequence of non-negative random variables $\{X_k, k = 1, 2, \dots\}$, if they are independent and the cdf of X_k is given by $F(a^{k-1}x)$ for $k = 1, 2, \dots$, where a is a positive constant, then $\{X_k, k = 1, 2, \dots\}$ is called a geometric process (GP).

Some variants of the GP can be found in Wu [25] and other models for imperfect repair can be found in Wu [26, 28].

1.3.2 *Two Widely Used Replacement Policies*

Preventive replacement can be worthwhile only if both the following two conditions hold:

- The failure intensity function of the item under maintenance is increasing; and
- The cost of failure replacement is greater than that of preventive replacement.

On a replacement policy, failures and repair may incur different types of costs, which include:

- cost of materials such as a new identical component;
- cost of labour and related overheads, and cost of associated administration; and
- cost of lost production in an average case.

1.3.2.1 Block Replacement Policy

With the block replacement policy, an item is replaced every t_0^* units of time. This implies that a replacement needs conducting although other replacement may have been done due to a failure prior to t_0^*, where t_0^* can be obtained by minimising the expected cost of the block replacement policy (BRP), which is given by

$$C(t_0) = \frac{c_r + c_m B(t_0)}{t_0}. \tag{1.2}$$

$B(t_0) = E[N(t_0)]$ is the expected number of replacements within time interval $(0, t)$. We have $B(t_0) = E[N(t_0)] = m(t)$, where $m(t)$ is given by

$$m(t) = F(t) + \int_0^t m(t-u) f(u) du. \tag{1.3}$$

Assuming $F(t)$ is a Weibull distribution, it is not straightforward to compute $m(t)$. The methods of obtaining the approximate of $m(t)$ has been well developed in the literature (see Xie [29], Jiang [11], for example). Suppose it is desired to approximate $m(t)$ with $0 \le t \le T^*$. Partition $[0, T^*]$ into N subintervals: $0 = T_0 < T_1 < \ldots < T_N = T^*$. The following recursive approximations to $m(T_i)$ in Xie [29] may be used:

$$m(T_i) = \frac{F(T_i) + S_i - F(T_i - T_{i-\frac{1}{2}}) m(T_{i-1})}{1 - F(T_i - T_{i-\frac{1}{2}})}, 1 \le i \le N, \tag{1.4}$$

with $m(T_0) = 0$, where $S_1 = 0$,

$$S_i = \sum_{j=1}^{i-1} F(T_i - T_{j-\frac{1}{2}})(m(T_j) - m(T_{j-1})), 2 \le i \le N, \tag{1.5}$$

and

$$T_{i-\frac{1}{2}} = \frac{1}{2}(T_{i-1} + T_i).$$

1.3.2.2 Age Replacement Policy

With the age replacement policy, an item is replaced whenever its age after the last replacement, which can be a scheduled replacement or a replacement on failure, reaches the optimised value t_0^*. The value t_0^* is usually obtained by minimising the expected cost of the age replacement policy (ARP), which is given by

$$C(t_0) = \frac{c_r + c_m F(t_0)}{\int_0^{t_0} (1 - F(t) dt}. \tag{1.6}$$

Comparing the expected cost of the age replacement policy (ECARP) with that of the block replacement policy (ECBRP), we can find that optimising ECARP is easier than optimising ECBRP, due to the difficulty in computing the renewal function $m(t)$.

1.4 Optimisation of Maintenance Policies for Items with Observable Failure Progression

In some real-world examples, the deterioration process of an item is observable. For example, the deterioration process of a section of pavement can be observed by measuring the number of defects such as the diameter of potholes and the length of cracks on it. The observed data can then be used to fit a model, which can be a gamma process if the increment rate increases in one direction, a Wiener process if the increment rate increases in two directions, or a Markov chain if the failure progression is approximated by a discrete set of states.

Let $X(t)$ be the degradation level of a degradation process at time t. Below gives the definitions of the gamma process and the Wiener process, respectively.

1.4.1 Gamma Process

Suppose the following assumptions hold:

- $X(0) = 0$;
- Increments $\Delta X(t) = X(t + \Delta t) - X(t)$ are independent of t;
- $\Delta X(t)$ also follows a gamma distribution Gamma$(\alpha(t + \Delta t) - \alpha(t), \beta)$ whose shape parameter is $\alpha(t + \Delta t) - \alpha(t)$ and scale parameter is β.

$X(t)$ has probability distribution Gamma$(\alpha(t), \beta)$ with mean $\beta\alpha(t)$, variance $\beta^2\alpha(t)$, and its probability density function being given by

$$f(x; \alpha(t), \beta) = \frac{\beta^{-\alpha(t)}}{\Gamma(\alpha(t))} x^{\alpha(t)-1} e^{-x/\beta} 1_{\{x>0\}}, \tag{1.7}$$

where $\Gamma(.)$ is the gamma function.

Then $\{X(t), x > 0\}$ is a gamma process.

The gamma process has the characteristics that it can only be used in optimisation of maintenance policies for single-component systems, but not for multi-components systems or systems with multi-failure modes [18]. Existing literature has well described the characteristic of the gamma process with applications.

1.4.2 Wiener Process

A Wiener process $\{X(t), x > 0\}$ normally follows the following assumptions:

- $X(0) = 0$, which also means that $W(0) = 0$;
- $W(t)$ has independent increments which follows the normal distribution. For $0 < s < t$, $W(t - s) - W(s)$ follows $N(0, (t - s))$.

- $W(t)$ is continuous in t.

Then $W(\cdot)$ is a standard Wiener process.
Meanwhile,

$$X(t) = \mu t + \sigma W(t), \tag{1.8}$$

is said a Wiener process with drift coefficient μ and variance parameter σ^2.

1.4.3 *Maintenance Policy for Items Modelled by the Gamma Process or the Weiner Process*

To characterise the maintenance policy for an item modelled by the gamma process or the Weiner process, the distribution of the first hitting times of the process $\{X(t), t \geq 0\}$ should normally be obtained. Starting from $X(0) = 0$ and for a fixed degradation level L, the first hitting time σ_L is defined as the amount of time required for the process $\{X(t), t \geq 0\}$ to reach the degradation level L, that is,

$$\sigma_L = \inf(t > 0 : X(t) \geq L).$$

The distribution of σ_L is obtained as

$$F_{\sigma_L}(t) = \Pr(X(t) \geq L). \tag{1.9}$$

Essentially, $F_{\sigma_L}(t)$ in Eq. (1.9) defines the reliability of the item and can therefore be used in maintenance policy optimisation, for example, $F_{\sigma_L}(t)$ can replace $F(t)$ in Eq. (1.6) and an age replacement policy can be obtained.

1.5 Importance Measures

Importance measures in reliability mathematics gauge the importance of a component in a system and can be used to identify the weakest components, which can provide helpful information in supporting performance improvement activities. Since Birnbaum [4] introduced the component importance measure in 1969, enormous types of importance measures have been proposed. Figure 1.7 illustrates a taxonomy of importance measures based on the information, where the rectangles on the right side give some examples.

Importance measures of Type I. Due to the fact that the structures of nowadays engineered systems are becoming more complex, it is often challenging to obtain the reliability block diagram (i.e., structure) of a system. Fortunately, engineers normally know the critical components in a system. For example, the counterweight,

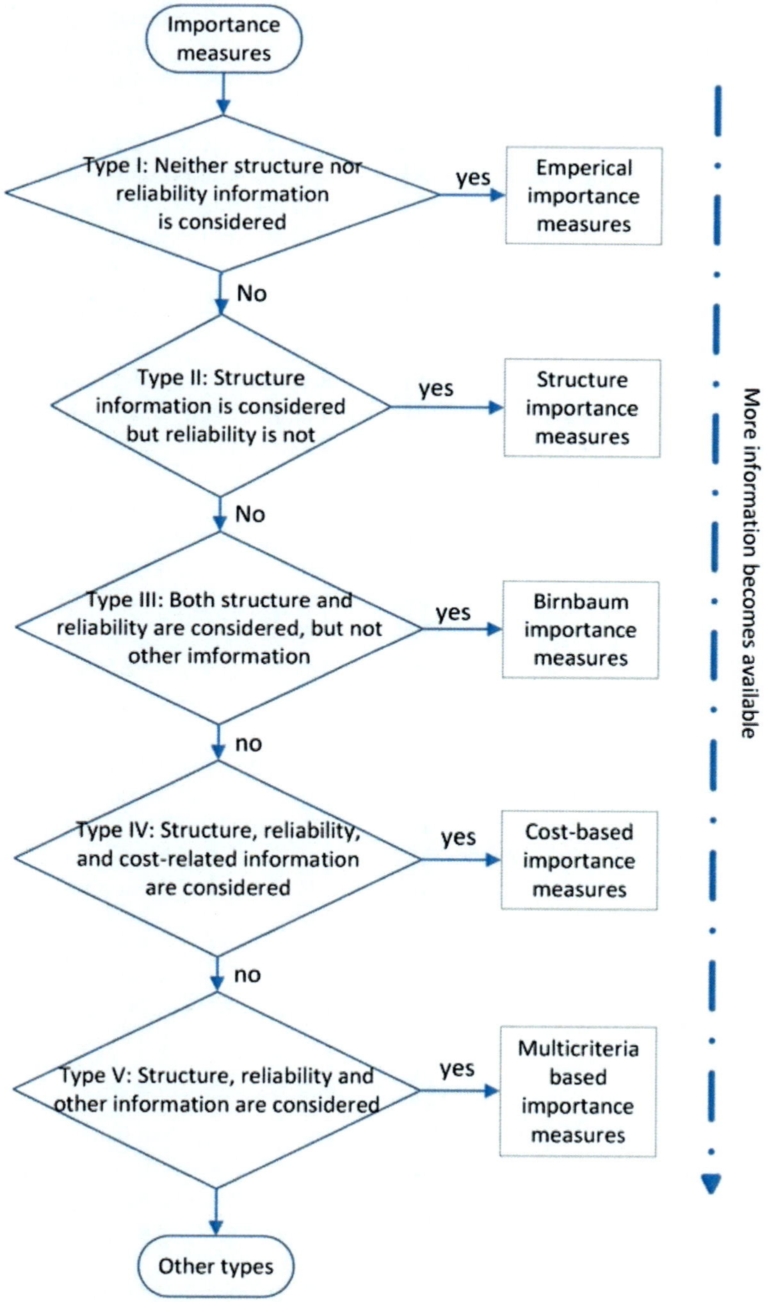

Fig. 1.7 A taxonomy of importance measures

the safety brake, and the speed governors are the most important components in a lift and should therefore be paid more attention from a safety perspective. Hence, Type I importance measures (or empirical importance measures) are defined when the structure of the system is not considered.

Importance measures of Type II. If the structure of a system is known and hence considered, importance measures such as the structure importance measures can defined. For example, supposing there are n components in a binary system, Lin [13] defines the structure importance of component i in the system as

$$\frac{\text{the number of events that component } i \text{ is critical to the system}}{2^{n-1}}.$$

As can be seen, the structure importance measure only uses the information of the structure of a system. That is, it can be obtained when the reliability of each component in a system is unknown. Wu [24] extended the structure importance measure from the binary system to the multistate system.

Importance measures of Type III. If the structure of a system and the reliability of each component in the system are available, an importance measure based on those information can be defined. For example, the Birnbaum importance measure falls in this category. Since this type of importance measures will be the focus of this book, we spend more space discussing it below.

The Birnbaum importance measure of component i for binary systems is as follows [4]:

$$I_i^{\mathrm{B}} = \Pr\{\Phi(\mathbf{X}) = 1 \,|\mathbf{X}_i = 1\} - \Pr\{\Phi(\mathbf{X}) = 1 \,|\mathbf{X}_i = 0\}\,, \tag{1.10}$$

where $\Phi(\mathbf{X})$ is the structural function of the system. The Birnbaum importance measure was then extended from the binary system to the multistate system, for which the importance of state m of component i's importance measure is defined by Griffith [9]:

$$I_{i_m}^{\mathrm{G}} = \sum_{j=1}^{M} \left(a_j - a_{j-1}\right) [\Pr(\Phi(m_i, X) \geq j) - \Pr(\Phi((m-1)_i, X) \geq j)]\,, \tag{1.11}$$

where the Griffith's importance $I_{i_m}^{G}$ can be interpreted as the change in the system performance when a component deteriorates from state m to state m-1. $0 = a_0 \leq a_1 \leq ...a_M$ represents the performance levels corresponding to the state space $\{0, 1, 2, \ldots, M\}$ of the system. $U = \sum_{j=1}^{M} a_j \Pr(\Phi(X) = j)$ represents the system performance function.

Other importance measures of Type III include the Natvig importance measure [15, 17], the composite importance measures [19, 20], etc.

The system reliability can also be decomposed by the following equation

$$\frac{dR_s(t)}{dt} = \sum_{i=1}^{n} \frac{\partial R_s(t)}{\partial R_i(t)} \frac{dR_i(t)}{dt} = -\sum_{i=1}^{n} f_i(t) \frac{\partial R_s(t)}{\partial R_i(t)}. \tag{1.12}$$

Based on the above equation, Si et al. [22] defined the the integrated importance of component i as

$$I_i^{im} = -f_i(t) \frac{\partial R_s(t)}{\partial R_i(t)}. \tag{1.13}$$

Similarly, following Griffith's importance, the integrated importance measure for the multistate system can be defined.

Importance measures of Type IV. The above-discussed importance measures merely use the information of a system itself, but not external information such as cost of maintenance. The reason we refer to information such as cost of maintenance as *external information* is that such information may vary over time or depend on different maintenance teams. For example, Wu and Coolen [27] introduces an importance measure that considers the cost of repair over the lifetime of the system.

Importance measures of Type V. While the cost-based importance measures include the external information as a decision criterion, more decision criteria can be considered in an importance measure. For instance, Almoghathawi et al. [2] propose a new approach to prioritising network components based on multiple importance measures using a multi criteria decision making method.

1.6 Resilience

Resilience refers to a system's ability to resist disturbances and recover quickly [14]. Resilience, as an extension of the concept of reliability, can reflect the ability of a system to be able to complete its pre-specified functions after attacks and destructions. It reflects not only the system's ability to resist destruction, but also the system's ability to recover after suffering a loss. The research of system resilience is essentially a study of system recovery and stability after the system failures. Currently, resilience is widely used in the field of systems engineering, such as power grids, transportation networks and other infrastructure systems, as well as financial markets, ecosystems and so on.

In the existing literature, the assessment methods of resilience can be categorized into two types: qualitative assessment as well as quantitative assessment [10]. Qualitative assessment is mainly based on expert judgment and subjective experience, and assesses the resilience level by describing and analysing the characteristics of the system. Quantitative assessment, on the other hand, measures and evaluates the resilience level of a system with the help of quantitative and mathematical models. This book focuses the research in quantitative assessment. Hosseini et al. [10] gave the following formula to define the resilience from a quantitative perspective:

$$R^{\text{resilience}}(t) = \frac{Q^{\text{recovery}}(t)}{Q^{\text{loss}}(t)}, \tag{1.14}$$

where $R^{\text{resilience}}(t)$ represents the resilience of the system at moment t, $Q^{\text{loss}}(t)$ represents the amount of loss in system performance at moment t, and $Q^{\text{recovery}}(t)$ represents the amount of recovery in system performance at moment t.

References

1. Armstrong MJ (1995) Joint reliability-importance of components. IEEE Trans Reliab 44(3):408–412
2. Almoghathawi Y, Barker K, Rocco CM, Nicholson CD (2017) A multi-criteria decision analysis approach for importance identification and ranking of network components. Reliab Eng Syst Saf 158:142–151
3. Barlow RE, Proschan F (1975) Importance of system components and fault tree events. Stoch Process Appl 3(2):153–173
4. Birnbaum L (1969) On the importance of different elements in a multi-element system. Multivar Anal 2:1–15
5. British Standard (2017) 13306: 2017: Maintenance—maintenance terminology. BSI Standards Publication
6. Brunelle RD, Kapur KC (1999) Review and classification of reliability measures for multistate and continuum models. IIE Trans 31(12):1171–1180
7. Doyen L, Gaudoin O (2004) Classes of imperfect repair models based on reduction of failure intensity or virtual age. Reliab Eng Syst Saf 84(1):45–56
8. Fussell J (1975) How to hand-calculate system reliability and safety characteristics. IEEE Trans Reliab 24(3):169–174
9. Griffith WS (1980) Multistate reliability models. J Appl Probab 17(3):735–744
10. Hosseini S, Barker K, Ramirez-Marquez JE (2016) A review of definitions and measures of system resilience. Reliab Eng Syst Saf 145:47–61
11. Jiang R (2010) A simple approximation for the renewal function with an increasing failure rate. Reliab Eng Syst Saf 95(9):963–969
12. Lam Y (1988) Geometric processes and replacement problem. Acta Mathematicae Applicatae Sinica 4:366–377
13. Lin FH (1998) Reliability importance of multicomponent systems. J Chinese Inst Ind Eng 15(2):103–111
14. Nan C, Sansavini G (2017) A quantitative method for assessing resilience of interdependent infrastructures. Reliab Eng Syst Saf 157:35–53
15. Natvig B (1985) New light on measures of importance of system components. Scand J Stat 43–54
16. Natvig B, Sørmo S, Holen AT, Høgåsen G (1986) Multistate reliability theory-a case study. Adv Appl Probab 18(4):921–932
17. Natvig B, Eide KA, Gåsemyr J, Huseby AB, Isaksen SL (2009) Simulation based analysis and an application to an offshore oil and gas production system of the natvig measures of component importance in repairable systems. Reliab Eng Syst Saf 94(10):1629–1638
18. van Noortwijk JM (2009) A survey of the application of gamma processes in maintenance. Reliab Eng Syst Saf 94(1):2–21
19. Ramirez-Marquez JE, Coit DW (2005) Composite importance measures for multi-state systems with multi-state components. IEEE Trans Reliab 54(3):517–529
20. Ramirez-Marquez JE, Coit DW (2007) Multi-state component criticality analysis for reliability improvement in multi-state systems. Reliab Eng Syst Saf 92(12):1608–1619

21. Ross SM (1996) Stochastic processes, 2nd edn. Wiley, New York
22. Si S, Dui H, Cai Z, Sun S (2012) The integrated importance measure of multi-state coherent systems for maintenance processes. IEEE Trans Reliab 61(2):266–273
23. Vesely W (1970) A time-dependent methodology for fault tree evaluation. Nuclear Eng Design 13(2):337–360
24. Wu S (2005) Joint importance of multistate systems. Comput Ind Eng 49(1):63–75
25. Wu S (2018) Doubly geometric processes and applications. J Oper Res Soc 69(1):66–77
26. Wu S (2019) A failure process model with the exponential smoothing of intensity functions. Eur J Oper Res 275(2):502–513
27. Wu S, Coolen FP (2013) A cost-based importance measure for system components: an extension of the Birnbaum importance. Eur J Oper Res 225(1):189–195
28. Wu S (2022) The double ratio geometric process for the analysis of recurrent events. Naval Res Logist (NRL) 69(3):484–495
29. Xie M (1989) On the solution of renewal-type integral equations. Commun Stat Simul Comput 18(1):281–293

Chapter 2
Importance Measures Informed Reliability Design

Abstract This chapter investigates the gradient calculation and geometric significance of important measures and proposes a new multi-criteria importance measure. This chapter studies component reliability importance measures with respect to the changes of the sequence of components in a system. It also studies joint importance measures for reliability design and for system reconfiguration, respectively.

Keywords Gradient · Multi-criteria importance measure · Integrated importance measure · Joint importance measure

As the complexity of systems continues increasing, to ensure a certain level of system reliability and security becomes increasingly challenging [5]. In order to achieve the reliability/performance of an engineered system, the weak components in a system need to be identified and then improved, for which reliability importance measures are useful tools. Research in this area is abundant. Levitin and Ben-Haim [22] proposed a protection importance measure to track bottlenecks. Si et al. [36, 37] proposed an importance measure that depicts and gauges the impact of component failures on the state distribution of a multistate system. Dui et al. [11] proposed an importance measure for analysing the impact of external factors on system performance. Wang et al. [40] introduced a PageRank algorithm to identify important components by considering the functional dependencies between components. To optimise the system structure, Dui et al. [12] investigated an importance measure for components whose ordering is changed during the life cycle of the system. Du et al. [10] studied importance measures of K-terminal networks under the assumption that side failures occur according to the branching process and discussed the Bayesian importance of the K-terminal networks. Do and Bérenguer [9] introduced a condition-based importance measure for multi-component systems and analysed the different levels of information and economic dependencies between components. Liu et al. [27] proposed an importance measure for components with continuous degradation (i.e., continuum components).

© The Author(s), under exclusive license to Springer Nature Switzerland AG 2024

H. Dui and S. Wu, *Importance-Informed Reliability Engineering*, Springer Series in Reliability Engineering, https://doi.org/10.1007/978-3-031-52455-4_2

2.1 Gradient Computations and Geometrical Meaning of Importance Measures

This section aims to investigate the gradient calculation and geometric significance of important measures for improving the performance of multistate systems. We make the following assumptions.

- The system under investigation is a coherent multistate system.
- All components are statistically independent.
- The state space of component i is $\{0, 1, 2, \ldots, M_i\}$ and that of the system is $\{0, 1, 2, \ldots, M\}$, where 0 represents the complete failure of the system or a component, M_i is the perfect functioning state of component i and M is the perfect state of the system. The performance of the system or a component increase from 0 to M or M_i.

In vector analysis, the gradient of a scalar field is the vector field pointing to the direction of the maximum growth rate of the scalar field, and its magnitude is the growth rate. Simply put, a spatial change can be represented by a slope, which represents the steepness and direction of the change. The gradient can also be used to measure the change of the scalar field in other directions, not only the direction of the maximum change.

In mathematics, the gradient of function $f(y_1, y_2, \ldots, y_n)$ is the vector field, as shown below.

$$\begin{aligned} \nabla f &= \mathrm{grad} f \\ &= \frac{\partial f}{\partial y_1}\overrightarrow{y_1} + \frac{\partial f}{\partial y_2}\overrightarrow{y_2} + \cdots + \frac{\partial f}{\partial y_n}\overrightarrow{y_n}, \end{aligned} \tag{2.1}$$

where $\overrightarrow{y_i}\,(i = 1, 2, \ldots, n)$ is the orthogonal unit vector pointing towards the coordinate direction of y_i increment.

To evaluate the overall impact of a component on system performance, one can extend the integrated importance of the state of a component to the integrated importance of a component, as shown below.

$$\begin{aligned} I_i^{im} &= \sum_{m=1}^{M_i} I_{i_m}^{im} \\ &= \sum_{m=1}^{M_i} P_{i_m} \lambda_{m,0}^{i} \frac{\partial U}{\partial P_{i_m}}. \end{aligned} \tag{2.2}$$

When a function is differentiable, the dot product of the gradient of the function and the given unit vector is equal to the directional derivative of the function in the direction of the unit vector. The gradient is a vector operation that operates on a scalar function to generate a vector, whose size is the maximum rate of change of the function at the gradient point and points in the direction of that maximum rate of change. Based on this, we obtain Theorem 2.1.

Theorem 2.1 $\nabla \Pr\{\Phi(\boldsymbol{X})=1\} = I_1^B\overrightarrow{\boldsymbol{P_{1_1}}} + I_2^B\overrightarrow{\boldsymbol{P_{2_1}}} + \cdots + I_n^B\overrightarrow{\boldsymbol{P_{n_1}}}$.

Proof If we can consider $\Pr\{\Phi(\boldsymbol{X})=1\}$ as a function relating to parameters $P_{1_1}, P_{2_1}, \ldots, P_{n1}$, that is

$$\Pr\{\Phi(\boldsymbol{X})=1\} = f(P_{1_1}, P_{2_1}, \ldots, P_{n_1}), \tag{2.3}$$

then the gradient of function $f(P_{1_1}, P_{2_1}, \ldots, P_{n_1})$ is

$$\begin{aligned}\nabla f &= \frac{\partial \Pr\{\Phi(\boldsymbol{X})=1\}}{\partial P_{1_1}}\overrightarrow{\boldsymbol{P_{1_1}}} + \frac{\partial \Pr\{\Phi(\boldsymbol{X})=1\}}{\partial P_{2_1}}\overrightarrow{\boldsymbol{P_{2_1}}} + \cdots + \frac{\partial \Pr\{\Phi(\boldsymbol{X})=1\}}{\partial P_{n_1}}\overrightarrow{\boldsymbol{P_{n_1}}} \\ &= I_1^{\mathrm{B}}\overrightarrow{\boldsymbol{P_{1_1}}} + I_2^{\mathrm{B}}\overrightarrow{\boldsymbol{P_{2_1}}} + \cdots + I_n^{\mathrm{B}}\overrightarrow{\boldsymbol{P_{n_1}}}.\end{aligned} \tag{2.4}$$

This proves the theorem. □

From Theorem 2.1, we can obtain that the gradient of function $f(P_{1_1}, P_{2_1}, \ldots, P_{n_1})$ is composed of the Birnbaum reliability importance of all components. The gradient at a certain point $(P_{1_1}, P_{2_1}, \ldots, P_{n_1})$ indicates the direction of the fastest increase in system reliability. The magnitude of the gradient will determine the speed at which system reliability increases in that direction.

Theorem 2.2 $\nabla U = \frac{\partial U}{\partial \rho_{i_1}}\overrightarrow{\boldsymbol{\rho_{i_1}}} + \frac{\partial U}{\partial \rho_{i_2}}\overrightarrow{\boldsymbol{\rho_{i_2}}} + \cdots + \frac{\partial U}{\partial \rho_{iM_i}}\overrightarrow{\boldsymbol{\rho_{iM_i}}}$.

Proof Considering $U = \sum_{j=1}^{M} a_j \Pr\{\Phi(\boldsymbol{X}) = j\}$ as a function with respect to the parameters $\rho_{i_1}, \rho_{i_2}, \ldots, \rho_{iM_i}$, and $\rho_{i_j} = \sum_{l=j}^{M} P_{i_l} = \sum_{l=j}^{M} P_r\{X_i = l\}$, we can obtain

$$U = \sum_{j=1}^{M}(a_j - a_{j-1})\Pr[(\Phi(0_i, \boldsymbol{X}) \geq j)] + I_i^{\mathrm{G}}\rho_i^T,\ \rho_i = \left(\rho_{i_1}, \rho_{i_2}, \ldots, \rho_{iM_i}\right), \tag{2.5}$$

where $I_i^{\mathrm{G}} = \left(I_{i_1}^G, I_{i_2}^G, \ldots, I_{iM_i}^G\right) = \left(\frac{\partial U}{\partial \rho_{i_1}}, \frac{\partial U}{\partial \rho_{i_2}}, \ldots, \frac{\partial U}{\partial \rho_{iM_i}}\right)$.

Then we have $\frac{\partial U}{\partial \rho_{im}} = I_{i_m}^{\mathrm{G}}$. Hence, we can obtain that

$$\begin{aligned}\nabla U &= \frac{\partial U}{\partial \rho_{i_1}}\overrightarrow{\boldsymbol{\rho_{i_1}}} + \frac{\partial U}{\partial \rho_{i_2}}\overrightarrow{\boldsymbol{\rho_{i_2}}} + \cdots + \frac{\partial U}{\partial \rho_{iM_i}}\overrightarrow{\boldsymbol{\rho_{iM_i}}} \\ &= I_{i_1}^G\overrightarrow{\boldsymbol{\rho_{i_1}}} + I_{i_2}^G\overrightarrow{\boldsymbol{\rho_{i_2}}} + I_{iM_i}^G\overrightarrow{\boldsymbol{\rho_{iM_i}}}.\end{aligned} \tag{2.6}$$

This proves the theorem. □

From Theorem 2.2, we can obtain that the gradient of function $\nabla \boldsymbol{U}$ is composed of the Griffith importance of all component states. When $U = \sum_{j=1}^{M} a_j \Pr(\Phi(\boldsymbol{X}) = j)$ is considered as a function with respect to the parameters $P_{i_1}, P_{i_2}, \ldots, P_{i_{M_i}}$, that is $U = f(P_{i_1}, P_{i_2}, \ldots, P_{i_{M_i}})$, we can obtain

$$
\begin{aligned}
\frac{\partial U}{\partial P_{i_m}} &= \sum_{j=1}^{M} \left(a_j - a_{j-1}\right) \left[\Pr\left(\Phi(m_i, \boldsymbol{X}) \geq j\right) - \Pr\left(\Phi(0_i, \boldsymbol{X}) \geq j\right) \right] \\
&= \sum_{j=1}^{M} a_j \left[\Pr\left(\Phi(m_i, \boldsymbol{X}) = j\right) - \Pr\left(\Phi(0_i, \boldsymbol{X}) = j\right) \right].
\end{aligned} \tag{2.7}
$$

In this case, we can obtain $\nabla \boldsymbol{U} = \frac{\partial U}{\partial P_{i_1}} \overrightarrow{\boldsymbol{P_{i_1}}} + \frac{\partial U}{\partial P_{i_2}} \overrightarrow{\boldsymbol{P_{i_2}}} + \cdots + \frac{\partial U}{\partial P_{i_{M_i}}} \overrightarrow{\boldsymbol{P_{i_{M_i}}}}$. So the gradient of function U is relating to the parameters.

Assuming the parameters $P_{i_m} = x_m$, in the rectangular coordinate system $\overrightarrow{\boldsymbol{x_1}}, \overrightarrow{\boldsymbol{x_2}}, \ldots, \overrightarrow{\boldsymbol{x_{M_i}}}$, we have $\nabla \boldsymbol{U} = \frac{\partial U}{\partial P_{i_1}} \overrightarrow{\boldsymbol{x_1}} + \frac{\partial U}{\partial P_{i_2}} \overrightarrow{\boldsymbol{x_2}} + \cdots + \frac{\partial U}{\partial P_{i_{M_i}}} \overrightarrow{\boldsymbol{x_{M_i}}}$.

Theorem 2.3 *Let $u_m = P_{i_m} \lambda^i_{m,0} x_m$, $m = 1, 2, \ldots, M_i$. Then the coordinate system $\overrightarrow{\boldsymbol{u_1}}, \overrightarrow{\boldsymbol{u_2}}, \ldots, \overrightarrow{\boldsymbol{u_{M_i}}}$ is orthogonal.*

Proof Since $u_m = P_{i_m} \lambda^i_{m,0} x_m$, $m = 1, 2, \ldots, M_i$, we have $x_m = \frac{u_m}{P_{i_m} \lambda^i_{m,0}}$. It is also expressed as

$$
\boldsymbol{x} = \left(x_1, x_2, \ldots, x_{M_i}\right) = \left(\frac{u_1}{P_{i_1} \lambda^i_{1,0}}, \frac{u_2}{P_{i_2} \lambda^i_{2,0}}, \ldots, \frac{u_{M_i}}{P_{i_{M_i}} \lambda^i_{M_i,0}} \right). \tag{2.8}
$$

We can hence obtain

$$
\begin{aligned}
\frac{\partial \boldsymbol{x}}{\partial \boldsymbol{u_1}} &= \left(\frac{1}{P_{i_1} \lambda^i_{1,0}}, 0, \ldots, 0 \right), \\
\frac{\partial \boldsymbol{x}}{\partial \boldsymbol{u_2}} &= \left(0, \frac{1}{P_{i_2} \lambda^i_{2,0}}, \ldots, 0 \right), \\
&\cdots \\
\frac{\partial \boldsymbol{x}}{\partial \boldsymbol{u_{M_i}}} &= \left(0, 0, \ldots, \frac{1}{P_{i_{M_i}} \lambda^i_{M_i,0}} \right).
\end{aligned} \tag{2.9}
$$

Then we have

$$
\frac{\partial \boldsymbol{x}}{\partial \boldsymbol{u_m}} \frac{\partial \boldsymbol{x}}{\partial \boldsymbol{u_k}} = 0, m \neq k, \quad \frac{\partial \boldsymbol{x}}{\partial \boldsymbol{u_m}} \frac{\partial \boldsymbol{x}}{\partial \boldsymbol{u_m}} = \left(\frac{1}{P_{i_1} \lambda^i_{1,0}} \right)^2, \tag{2.10}
$$

where the dot product is the sum of the products of the corresponding entries of two sequences of vector $\frac{\partial x}{\partial u_m}$ and vector $\frac{\partial x}{\partial u_k}$.

As such, the coordinate system $\overrightarrow{u_1}, \overrightarrow{u_2}, \ldots, \overrightarrow{u_{M_i}}$ is orthogonal. □

From Theorem 2.3, we have

$$\begin{aligned}\nabla U &= \frac{\partial U}{\partial u_1}\frac{\partial u_1}{\partial x_1}\overrightarrow{u_1} + \frac{\partial U}{\partial u_2}\frac{\partial u_2}{\partial x_2}\overrightarrow{u_2} + \cdots + \frac{\partial U}{\partial u_{M_i}}\frac{\partial u_{M_i}}{\partial x_{M_i}}\overrightarrow{u_{M_i}} \\ &= P_{i_1}\lambda^i_{1,0}\frac{\partial U}{\partial u_1}\overrightarrow{u_1} + P_{i_2}\lambda^i_{2,0}\frac{\partial U}{\partial u_2}\overrightarrow{u_2} + \cdots + P_{i_{M_i}}\lambda^i_{M_i,0}\frac{\partial U}{\partial u_{M_i}}\overrightarrow{u_{M_i}}.\end{aligned} \tag{2.11}$$

Given a vector $\overrightarrow{v} = P_{i_1}\lambda^i_{1,0}\overrightarrow{x_1} + P_{i_2}\lambda^i_{2,0}\overrightarrow{x_2} + \cdots + P_{i_{M_i}}\lambda^i_{M_i,0}\overrightarrow{x_{M_i}}$, based on Eq. (2.11), we can obtain

$$\begin{aligned}I_i^{im} &= \sum_{m=1}^{M_i} P_{i_m}\lambda^i_{m,0}\frac{\partial U}{\partial P_{i_m}} \\ &= \left(P_{i_1}\lambda^i_{1,0}, P_{i_2}\lambda^i_{2,0}, \ldots, P_{i_{M_i}}\lambda^i_{M_i,0}\right)\left(\frac{\partial U}{\partial P_{i_1}}, \frac{\partial U}{\partial P_{i_2}}, \ldots, \frac{\partial U}{\partial P_{i_{M_i}}}\right)^T \\ &= \overrightarrow{v}\left(\frac{\partial U}{\partial P_{i_1}}\overrightarrow{P_{i_1}} + \frac{\partial U}{\partial P_{i_2}}\overrightarrow{P_{i_2}} + \cdots + \frac{\partial U}{\partial P_{i_{M_i}}}\overrightarrow{P_{i_{M_i}}}\right).\end{aligned} \tag{2.12}$$

Equation (2.12) suggests that the integrated importance of component i is the dot product of $\overrightarrow{v}$ and the gradient of function U relating to the parameters $P_{i_1}, P_{i_2}, \ldots, P_{i_{M_i}}$.

The structural function of a series system composed of n components is $\Phi(X) = \min_{1\le k\le n}\{X_k\}$. The state space of each component is $\{0, 1, 2, \ldots, M\}$ and all the components are independent of each other.

In a binary series system,

$$\Pr\{\Phi(X) = 1\} = \sum_{i=1}^{n} P_{i_1}. \tag{2.13}$$

Then we can obtain

$$\frac{\partial \Pr\{\Phi(X) = 1\}}{\partial P_{i_1}} = \sum_{k=1,k\neq i}^{n} P_{k1}. \tag{2.14}$$

Hence, we have

$$\nabla \mathbf{Pr}\left\{\mathbf{\Phi}(X)=\mathbf{1}\right\}=\sum_{k=2}^{n} P_{k1}\overrightarrow{\boldsymbol{P}_{\mathbf{1}_1}}+\sum_{k=1,k\neq 2}^{n} P_{k1}\overrightarrow{\boldsymbol{P}_{\mathbf{2}_1}}+\cdots+\sum_{k=1}^{n-1} P_{k1}\overrightarrow{\boldsymbol{P}_{\boldsymbol{n}_1}}. \tag{2.15}$$

In a multistate series system, we have

$$\begin{aligned}
&\sum_{j=1}^{M}\left(a_j-a_{j-1}\right)\Pr\left(\Phi(m_i,\boldsymbol{X})\geq j\right)\\
&=\sum_{j=1}^{m}\left(a_j-a_{j-1}\right)\Pr\left(\Phi(m_i,\boldsymbol{X})\geq j\right)\\
&\quad+\sum_{j=1+m}^{M}\left(a_j-a_{j-1}\right)\Pr\left(\Phi(m_i,\boldsymbol{X})\geq j\right)\\
&=\sum_{j=1}^{m}\left(a_j-a_{j-1}\right)\Pr\left(\Phi(m_i,\boldsymbol{X})\geq j\right)\\
&=\sum_{j=1}^{m}\left(a_j-a_{j-1}\right)\Pr\left\{X_1\geq j,\ldots,X_{i-1}\geq j,m\geq j,X_{1+i}\geq j,\ldots,X_n\geq j\right\}\\
&=\sum_{j=1}^{m}\left(a_j-a_{j-1}\right)\sum_{k=1,k\neq i}^{n}\rho_{kj}.
\end{aligned} \tag{2.16}$$

Then we can obtain

$$\begin{aligned}
&\sum_{j=1}^{M}\left(a_j-a_{j-1}\right)\left[\Pr\left(\Phi(m_i,\boldsymbol{X})\geq j\right)-\Pr\left(\Phi((m-1)_i,\boldsymbol{X})\geq j\right)\right]\\
&=\sum_{j=1}^{m}\left(a_j-a_{j-1}\right)\sum_{k=1,k\neq i}^{n}\rho_{kj}-\sum_{j=1}^{m-1}\left(a_j-a_{j-1}\right)\sum_{k=1,k\neq i}^{n}\rho_{kj}\\
&=\left(a_m-a_{m-1}\right)\sum_{k=1,k\neq i}^{n}\rho_{km}.
\end{aligned} \tag{2.17}$$

Then

$$\begin{aligned}
\nabla\boldsymbol{U}=&\left(a_1-a_0\right)\sum_{k=1,k\neq i}^{n}\rho_{k1}\overrightarrow{\boldsymbol{\rho}_{\boldsymbol{i}_1}}+\left(a_2-a_1\right)\sum_{k=1,k\neq i}^{n}\rho_{k2}\overrightarrow{\boldsymbol{\rho}_{\boldsymbol{i}_2}}+\ldots\\
&+\left(a_{M_i}-a_{M_i-1}\right)\sum_{k=1,k\neq i}^{n}\rho_{kM_i}\overrightarrow{\boldsymbol{\rho}_{\boldsymbol{i}_{M_i}}}.
\end{aligned} \tag{2.18}$$

For a series system,

$$I_{i_m}^{im} = P_{i_m}\lambda_{m,0}^{i}\sum_{j=1}^{m}\left(a_j - a_{j-1}\right)\sum_{k=1,k\neq i}^{n}\rho_{kj}. \tag{2.19}$$

According to Eq. (2.6), we have

$$\frac{\partial U}{\partial P_{i_m}} = \sum_{j=1}^{m}\left(a_j - a_{j-1}\right)\sum_{k=1,k\neq i}^{n}\rho_{kj}. \tag{2.20}$$

Hence, we can obtain

$$\begin{aligned}\nabla \boldsymbol{U} =&(a_1 - a_0)\sum_{k=1,k\neq i}^{n}\rho_{k1}\overrightarrow{\boldsymbol{P}_{i_1}}\\&+\sum_{j=1}^{2}(a_j - a_{j-1})\sum_{k=1,k\neq i}^{n}\rho_{kj}\overrightarrow{\boldsymbol{P}_{i_2}} + \ldots\\&+\sum_{j=1}^{M_i}(a_j - a_{j-1})\sum_{k=1,k\neq i}^{n}\rho_{kj}\overrightarrow{\boldsymbol{P}_{i_{M_i}}}.\end{aligned} \tag{2.21}$$

The structural function of a parallel system composed of n components is $\Phi(\boldsymbol{X}) = \max\limits_{1\leq k\leq n}\{X_k\}$. The state space of each component is $\{0, 1, 2, \ldots, M\}$ and all the components are independent of each other.

In a binary parallel system, $\Pr\{\Phi(\boldsymbol{X}) = 1\} = 1 - \sum_{i=1}^{n}(1 - P_{i1})$. Then we can obtain $\frac{\partial \Pr\{\Phi(\boldsymbol{X})=1\}}{\partial P_{i_1}} = \sum_{k=1,k\neq i}^{n}(1 - P_{k1})$. Hence, we have $\nabla\Pr\{\Phi(\boldsymbol{X}) = 1\} = \sum_{k=2}^{n}(1 - P_{k1})\overrightarrow{\boldsymbol{P}_{1_1}} + \sum_{k=1,k\neq 2}^{n}(1 - P_{k1})\overrightarrow{\boldsymbol{P}_{2_1}} + \cdots + \sum_{k=1}^{n-1}(1 - P_{k1})\overrightarrow{\boldsymbol{P}_{n_1}}$.

For a parallel system, we have

$$\begin{aligned}&\sum_{j=1}^{M}\left(a_j - a_{j-1}\right)\Pr\left(\Phi(m_i, \boldsymbol{X}) < j\right)\\&= \sum_{j=1}^{m}\left(a_j - a_{j-1}\right)\Pr\left(\Phi(m_i, \boldsymbol{X}) < j\right) + \sum_{j=1+m}^{M}\left(a_j - a_{j-1}\right)\Pr\left(\Phi(m_i, \boldsymbol{X}) < j\right)\\&= \sum_{j=1+m}^{M}\left(a_j - a_{j-1}\right)\Pr\left(\Phi(m_i, \boldsymbol{X}) < j\right) = \sum_{j=1+m}^{M}\left(a_j - a_{j-1}\right)\sum_{k=1,k\neq i}^{n}(1 - \rho_{kj}).\end{aligned} \tag{2.22}$$

So, we can obtain

$$\sum_{j=1}^{M}\left(a_j - a_{j-1}\right)\Pr\left(\Phi(m_i, \boldsymbol{X}) \geq j\right) - \Pr\left(\Phi((m-1)_i, \boldsymbol{X}) \geq j\right)]$$

$$= \sum_{j=1}^{M}\left(a_j - a_{j-1}\right)\Pr\left(\Phi((m-1)_i, \boldsymbol{X}) < j\right) - \Pr\left(\Phi(m_i, \boldsymbol{X}) < j\right)]$$

$$= \sum_{j=1}^{M}\left(a_j - a_{j-1}\right)\sum_{k=1,k\neq i}^{n}(1-\rho_{kj}) - \sum_{j=m+1}^{M}\left(a_j - a_{j-1}\right)\sum_{k=1,k\neq i}^{n}(1-\rho_{kj})$$

$$= (a_m - a_{m-1})\sum_{k=1,k\neq i}^{n}(1-\rho_{km}). \tag{2.23}$$

Then

$$\nabla \boldsymbol{U} = (a_1 - a_0)\sum_{k=1,k\neq i}^{n}(1-\rho_{k1})\,\overrightarrow{\boldsymbol{\rho_{i_1}}} + (a_2 - a_1)\sum_{k=1,k\neq i}^{n}(1-\rho_{k2})\,\overrightarrow{\boldsymbol{\rho_{i_2}}} + \ldots$$
$$+ \left(a_{M_i} - a_{M_i-1}\right)\sum_{k=1,k\neq i}^{n}\left(1-\rho_{kM_i}\right)\overrightarrow{\boldsymbol{\rho_{i_{M_i}}}}. \tag{2.24}$$

In a series system,

$$I_{i_m}^{im} = P_{i_m}\lambda_{m,0}^{i}\sum_{j=1}^{m}\left(a_j - a_{j-1}\right)\sum_{k=1,k\neq i}^{n}(1-\rho_{kj}). \tag{2.25}$$

Hence, we obtain

$$\frac{\partial U}{\partial P_{i_m}} = \sum_{j=1}^{m}\left(a_j - a_{j-1}\right)\sum_{k=1,k\neq i}^{n}(1-\rho_{kj}), \tag{2.26}$$

and

$$\nabla \boldsymbol{U} = (a_1 - a_0)\sum_{k=1,k\neq i}^{n}(1-\rho_{k1})\,\overrightarrow{\boldsymbol{P_{i_1}}} + \sum_{j=1}^{2}\left(a_j - a_{j-1}\right)\sum_{k=1,k\neq i}^{n}\left(1-\rho_{kj}\right)\overrightarrow{\boldsymbol{P_{i_2}}} + \ldots$$
$$+ \sum_{j=1}^{M_i}\left(a_j - a_{j-1}\right)\sum_{k=1,k\neq i}^{n}\left(1-\rho_{kj}\right)\overrightarrow{\boldsymbol{P_{i_{M_i}}}}. \tag{2.27}$$

Remark 2.1 We can obtain the following remarks.

- In the geometric space, $\Pr\{\Phi(X) = 1\} = f(P_{1_1}, P_{2_1}, \ldots, P_{n1}) = R$ represents a curve surface. The vector $(I_1^{\mathrm{B}}, I_2^{\mathrm{B}}, \ldots, I_n^{\mathrm{B}})$ is the normal to the iso-surface $f(P_{1_1}, P_{2_1}, \ldots, P_{n_1}) = R$. This direction implies the fastest increase in system reliability.
- In the geometric space, $U = f(\rho_{1_1}, \rho_{2_1}, \ldots, \rho_{i_{M_i}}) = u$ represents a curve surface. The vector $(I_1^G(i), I_2^G(i), \ldots, I_{M_i}^G(i))$ is the normal to the iso-surface $f(\rho_{i_1}, \rho_{i_2}, \ldots, \rho_{i_{M_i}}) = u$. This direction implies the fastest increase in system reliability.
- In the geometric space, $U = f(P_{i_1}, P_{i_2}, \ldots, P_{i_{M_i}}) = u$ represents a curve surface. The vector $(\frac{\partial U}{\partial P_{i_1}}, \frac{\partial U}{\partial P_{i_2}}, \ldots, \frac{\partial U}{\partial P_{i_{M_i}}})$ is the normal to the iso-surface $f(P_{i_1}, P_{i_2}, \ldots, P_{i_{M_i}}) = u$. This direction implies the fastest increase in system reliability. The integrated importance of component i ($\mathrm{I}_i^{\mathrm{im}}$) at point $(P_{i_1}, P_{i_2}, \ldots, P_{i_{M_i}})$ is the product of the projected length of the vector $\overrightarrow{\boldsymbol{v}}$ in the gradient $\frac{\partial U}{\partial P_{i_1}}\overrightarrow{\boldsymbol{P}_{i_1}} + \frac{\partial U}{\partial P_{i_2}}\overrightarrow{\boldsymbol{P}_{i_2}} + \cdots + \frac{\partial U}{\partial P_{i_{M_i}}}\overrightarrow{\boldsymbol{P}_{i_{M_i}}}$ and the magnitude of the gradient $\frac{\partial U}{\partial P_{i_1}}\overrightarrow{\boldsymbol{P}_{i_1}} + \frac{\partial U}{\partial P_{i_2}}\overrightarrow{\boldsymbol{P}_{i_2}} + \cdots + \frac{\partial U}{\partial P_{i_{M_i}}}\overrightarrow{\boldsymbol{P}_{i_{M_i}}}$ at point $(P_{i_1}, P_{i_2}, \ldots, P_{i_{M_i}})$. The projected length of the vector $\overrightarrow{\boldsymbol{v}}$ in the gradient $\frac{\partial U}{\partial P_{i_1}}\overrightarrow{\boldsymbol{P}_{i_1}} + \frac{\partial U}{\partial P_{i_2}}\overrightarrow{\boldsymbol{P}_{i_2}} + \cdots + \frac{\partial U}{\partial P_{i_{M_i}}}\overrightarrow{\boldsymbol{P}_{i_{M_i}}}$ means the characteristic of transition rates among different component states.

Numerical examples are as follows.

Example 2.1 In a binary series system composed of three components,

$$\Pr\{\Phi(X) = 1\} = f\left(P_{1_1}, P_{2_1}, P_{3_1}\right) = P_{1_1} P_{2_1} P_{3_1}. \tag{2.28}$$

Assuming $P_{1_1} = 0.5$, $P_{2_1} = 0.6$, $P_{3_1} = 0.7$ and $f\left(P_{1_1}, P_{2_1}, P_{3_1}\right) = P_{1_1} P_{2_1} P_{3_1} = 0.21$, the gradient of $f\left(P_{1_1}, P_{2_1}, P_{3_1}\right)$ is

$$\begin{aligned}\nabla f\left(P_{1_1}, P_{2_1}, P_{3_1}\right) &= I_1^{\mathrm{B}}\overrightarrow{\boldsymbol{P}_{1_1}} + I_2^{\mathrm{B}}\overrightarrow{\boldsymbol{P}_{2_1}} + I_3^{\mathrm{B}}\overrightarrow{\boldsymbol{P}_{3_1}} \\ &= 0.42\overrightarrow{\boldsymbol{P}_{1_1}} + 0.35\overrightarrow{\boldsymbol{P}_{2_1}} + 0.3\overrightarrow{\boldsymbol{P}_{3_1}}.\end{aligned} \tag{2.29}$$

Similarly, in a binary parallel system composed of three components, assuming $P_{1_1} = 0.5$, $P_{2_1} = 0.6$, $P_{3_1} = 0.7$ and $f(P_{1_1}, P_{2_1}, P_{3_1}) = 1 - \sum_{i=1}^{3}(1 - P_{i_1}) = 0.94$, the gradient of $f(P_{1_1}, P_{2_1}, P_{3_1})$ is $\nabla f\left(P_{1_1}, P_{2_1}, P_{3_1}\right) = 0.12\overrightarrow{\boldsymbol{P}_{1_1}} + 0.15\overrightarrow{\boldsymbol{P}_{2_1}} + 0.2\overrightarrow{\boldsymbol{P}_{3_1}}$.

In a system composed of two components, assuming the system and components all have four states (0,1,2,3) in a series system, we have

$$\begin{aligned}U &= \sum_{j=1}^{3} a_j \Pr\{\Phi(X_1, X_2) = j\} \\ &= f\left(\rho_{1_1}, \rho_{1_2}, \rho_{1_3}\right) = a_1 \rho_{2_1} \rho_{1_1} + (a_2 - a_1)\rho_{2_2}\rho_{1_2} + (a_3 - a_2)\rho_{2_3}\rho_{1_3}.\end{aligned} \tag{2.30}$$

Assuming $a_1 = 150,\ a_2 = 200,\ a_3 = 250,\ P_{2_1} = 0.3,\ P_{2_2} = 0.4,\ P_{2_3} = 0.2$ and the iso-surface is

$$\begin{aligned} f\left(\rho_{1_1}, \rho_{1_2}, \rho_{1_3}\right) &= a_1 \rho_{2_1} \rho_{1_1} + (a_2 - a_1)\, \rho_{2_2} \rho_{1_2} + (a_3 - a_2)\, \rho_{2_3} \rho_{1_3} \\ &= 135\overrightarrow{\boldsymbol{\rho_{1_1}}} + 30\overrightarrow{\boldsymbol{\rho_{1_2}}} + 10\overrightarrow{\boldsymbol{\rho_{1_3}}} \\ &= 134.5. \end{aligned} \tag{2.31}$$

So the gradient of $f\left(\rho_{1_1}, \rho_{1_2}, \rho_{1_3}\right)$ is $\nabla f\left(\rho_{1_1}, \rho_{1_2}, \rho_{1_3}\right) = 135\overrightarrow{\boldsymbol{\rho_{1_1}}} + 30\overrightarrow{\boldsymbol{\rho_{1_2}}} + 10\overrightarrow{\boldsymbol{\rho_{1_3}}}$. Similarly, in a parallel system, we have

$$\begin{aligned} U = f\left(\rho_{1_1}, \rho_{1_2}, \rho_{1_3}\right) &= a_1 \rho_{2_1} P_{10} + (a_2 - a_1)\, \rho_{2_2} P_{10} + (a_3 - a_2)\, \rho_{2_3} P_{10} \\ &+ a_1(1 - \rho_{2_1})\rho_{1_1} + (a_2 - a_1)\,(1 - \rho_{2_2})\rho_{1_2} + (a_3 - a_2)\,(1 - \rho_{2_3})\rho_{1_3}. \end{aligned} \tag{2.32}$$

Assuming $a_1 = 150,\ a_2 = 200,\ a_3 = 250,\ P_{2_1} = 0.3,\ P_{2_2} = 0.4,\ P_{2_3} = 0.2$, $P_{1_0} = 0.1$ and the iso-surface is $f\left(\rho_{1_1}, \rho_{1_2}, \rho_{1_3}\right) = 15\overrightarrow{\boldsymbol{\rho_{1_1}}} + 20\overrightarrow{\boldsymbol{\rho_{1_2}}} + 40\overrightarrow{\boldsymbol{\rho_{1_3}}} + 17.5 = 43$, the gradient of $f\left(\rho_{1_1}, \rho_{1_2}, \rho_{1_3}\right)$ is $\nabla f\left(\rho_{1_1}, \rho_{1_2}, \rho_{1_3}\right) = 15\overrightarrow{\boldsymbol{\rho_{1_1}}} + 20\overrightarrow{\boldsymbol{\rho_{1_2}}} + 40\overrightarrow{\boldsymbol{\rho_{1_3}}}$.

In a system consisting of 2 components, we assume the system and components all have 4 states (0,1,2,3). Assuming $P_{1_1} = 0.5,\ P_{1_2} = 0.3,\ P_{1_3} = 0.1,\ \lambda_{1,0}^1 = 3.5,\ \lambda_{2,0}^1 = 2.7,\ \lambda_{3,0}^1 = 1.9$, we can obtain the vector $\overrightarrow{\boldsymbol{v}} = P_{1_1}\lambda_{1,0}^1\overrightarrow{\boldsymbol{P_{1_1}}} + P_{1_2}\lambda_{2,0}^1\overrightarrow{\boldsymbol{P_{1_2}}} + P_{1_3}\lambda_{3,0}^1\overrightarrow{\boldsymbol{P_{1_3}}} = 1.75\overrightarrow{\boldsymbol{P_{1_1}}} + 0.81\overrightarrow{\boldsymbol{P_{1_2}}} + 0.19\overrightarrow{\boldsymbol{P_{1_3}}}$. In a series system, we have

$$\begin{aligned} U = f\left(P_{1_1}, P_{1_2}, P_{1_3}\right) &= a_1(P_{21} + P_{2_2} + P_{2_3})P_{1_1} + \left[a_2\left(P_{2_2} + P_{2_3}\right) + a_1 P_{2_1}\right] P_{1_2} \\ &+ P_{1_3} + (a_1 P_{2_1} + a_2 P_{2_2} + a_3 P_{2_3}). \end{aligned} \tag{2.33}$$

Assuming $a_1 = 150, a_2 = 200, a_3 = 250, P_{2_1} = 0.3, P_{2_2} = 0.4, P_{2_3} = 0.2$ and the iso-surface is $f\left(P_{1_1}, P_{1_2}, P_{1_3}\right) = 135P_{1_1} + 165P_{12} + 175P_{1_3} = 134.5$, the gradient of $f(P_{1_1}, P_{1_2}, P_{1_3})$ is $\nabla f\left(P_{1_1}, P_{1_2}, P_{1_3}\right) = 135\overrightarrow{\boldsymbol{P_{1_1}}} + 165\overrightarrow{\boldsymbol{P_{1_2}}} + 175\overrightarrow{\boldsymbol{P_{1_3}}}$. Similarly, in a parallel system, we have

$$\begin{aligned} U = f\left(P_{1_1}, P_{1_2}, P_{1_3}\right) &= a_1 P_{20} P_{1_1} + \left[a_2\left(P_{20} + P_{2_1}\right) - a_1 P_{2_1}\right] P_{1_2} \\ &+ \left(a_3 - a_2 P_{2_3}\right) P_{1_3} + a_1 P_{2_1} + a_2 P_{2_2} + a_3 P_{2_3}. \end{aligned} \tag{2.34}$$

Assuming $a_1 = 150, a_2 = 200, a_3 = 250, P_{2_1} = 0.3, P_{2_2} = 0.4, P_{2_3} = 0.2$ and the iso-surface is $f\left(P_{1_1}, P_{1_2}, P_{1_3}\right) = 175 + 15P_{1_1} + 35P_{1_2} + 200P_{1_3} = 213$, the gradient of $f(P_{1_1}, P_{1_2}, P_{1_3})$ is $\nabla f\left(P_{1_1}, P_{1_2}, P_{1_3}\right) = 15\overrightarrow{\boldsymbol{P_{1_1}}} + 35\overrightarrow{\boldsymbol{P_{1_2}}} + 200\overrightarrow{\boldsymbol{P_{1_3}}}$.

2.1.1 A New Multi-criteria Importance Measure Oriented to Reliability Improvement

When evaluating the importance of components, multiple criteria need to be considered comprehensively. The challenge lies in combining multiple standards while considering their correlation and obtaining importance measure that represents the comprehensive attributes of components in the system.

Importance measures based on a single standard can no longer reflect the true importance of system components. A multi-criteria importance measure (MCIM) is needed.

An engineered system is defined by $M(F, X)$, where $F = \{f_g | g = 1, 2, \ldots, m\}$ denotes the set of m criteria to be considered and f_g denotes the gth criterion. $X = \{x_k | k = 1, 2, \ldots, n\}$ denotes the set of system components, and x_k denotes the kth component. The MCIM for component x_k is denoted as I_k^{mc}. This section aims to define MCIM by considering the correlation between different standards, and demonstrates the application of the proposed MCIM in quickly and effectively identifying weak components in complex engineered systems.

A system contribution is an overall representation of all components' attribute under the same criteria. Therefore, the relationship between the contribution of a system under the f_{gth} criterion $v_g(\boldsymbol{X})$ and the contribution of x_k under the f_{gth} criterion $v_g(x_k)$ can be represented by a function, as shown in Eq. (2.35).

$$v_g(\boldsymbol{X}) = f_g(v_g(x_1), v_g(x_2), \ldots, v_g(x_n)). \tag{2.35}$$

Assume that these components are independent of each other. Then, under the given criteria, the impact of $v_g(x_k)$ on $v_g(\boldsymbol{X})$ can be expressed in terms of sensitivity as

$$J_{g^k} = \frac{\partial v_g(\boldsymbol{X})}{\partial v_g(x_k)}. \tag{2.36}$$

Under the criterion f_g, the attribute sensitivity of each component forms a component attribute gradient vector, denoted by ∇f_g.

$$\nabla f_g = \left[\frac{\partial v_g(\boldsymbol{X})}{\partial v_g(x_1)}, \frac{\partial v_g(\boldsymbol{X})}{\partial v_g(x_2)}, \ldots, \frac{\partial v_g(\boldsymbol{X})}{\partial v_g(x_n)}\right]. \tag{2.37}$$

The gradient vectors under all m criteria form an $m \times n$ Jacobian matrix J.

$$J = \frac{\partial(v_1(X), \ldots, v_m(X))}{\partial(v_g(x_1), \ldots, v_g(x_n))} = \begin{bmatrix} \frac{\partial v_1(X)}{\partial v_1(x_1)} & \cdots & \frac{\partial v_1(X)}{\partial v_1(x_n)} \\ \vdots & \ddots & \vdots \\ \frac{\partial v_m(X)}{\partial v_m(x_1)} & \cdots & \frac{\partial v_m(X)}{\partial v_m(x_n)} \end{bmatrix}. \tag{2.38}$$

J_{gk} is a scalar, so we do not need to consider the different units between the criteria. The distribution of J_{gk} varies between the different criteria, which may cause bias in multi-criteria decision making. However, this bias can be removed by standardising relevant variables. The max-min method is used to normalise the elements of the matrix J by rows and to obtain a normalised matrix J'. Depending on the condition of use, the criteria are divided into benefit criteria and cost criteria.

Let $\max J_{gk} = \max\limits_{1\le k\le n}\{J_{gk}\}$ and $\min J_{gk} = \min\limits_{1\le k\le n}\{J_{gk}\}$. J'_{gk} denotes the elements of the gth row and kth column of the normalised matrix J', respectively, where the normalised matrix J' is obtained as shown below

$$J' = \begin{bmatrix} J'_{11} & \cdots & J'_{1n} \\ \vdots & \ddots & \vdots \\ J'_{m1} & \cdots & J'_{mn} \end{bmatrix}, \tag{2.39}$$

and

$$J'_{gk} = \begin{cases} \frac{J_{gk}-\min J_{gk}}{\max J_{gk}-\min J_{gk}}, & \text{for benefit criteria} \\ \frac{\max J_{gk}-J_{gk}}{\max J_{gk}-\min J_{gk}}, & \text{for cost criteria} \end{cases}. \tag{2.40}$$

The $J'_{gk} \in [0, 1]$ and J'_{gk} tends to 0, which implies that the component is a weak component.

Then the matrix J' is split into columns to obtain the attribute characteristic vector $\boldsymbol{\phi}(\boldsymbol{x_k})$ for each component x_k.

$$\boldsymbol{\phi}(\boldsymbol{x_k}) = [J'_{1k}\ J'_{2k} \ldots\ J'_{mk}]^T, k = 1, 2, \ldots, n. \tag{2.41}$$

After normalisation, the optimal value of J'_{gk} for each criterion is 1. Therefore, we propose an ideal feature vector $\boldsymbol{B}$.

$$\boldsymbol{B}^T = [\overbrace{1\ 1\ \ldots\ 1}^{m}]. \tag{2.42}$$

The greater the difference between $\phi(x_k)$ and B of a component is, the weaker the component is. The multi-criteria importance I_k^{mc} can be quantified by the Euclidean distance, as shown in Eq. (2.43).

$$I_k^{mc} = d(\boldsymbol{B}, \boldsymbol{\phi}(\boldsymbol{x_k})) = \sqrt{\sum_{g=1}^{m}(1 - J'_{gk})^2}. \tag{2.43}$$

Equation (2.43) is based on the assumption that m criteria are independent of each other. However, in practice, some standards are interrelated or conflicting.

The correlation coefficient between criterion f_i and criterion f_j is represented by α_{ij} [1].

$$\alpha_{ij} = \frac{\sum_{k=1}^{n}(J'_{ik} - \overline{J'_i})(J'_{jk} - \overline{J'_j})}{\sqrt{\sum_{k=1}^{n}(J'_{ik} - \overline{J'_i})^2 \sum_{k=1}^{n}(J'_{jk} - \overline{J'_j})^2}}, \tag{2.44}$$

where $\overline{J'_g} = \frac{1}{n}\sum_{k=1}^{n} J'_{gk}$.

C_g denotes the conflicts of criterion f_g with other criteria, which can be expressed as

$$C_g = \sum_{g=1}^{m}(1 - \alpha_{ig}). \tag{2.45}$$

Let S_g denote the standard deviation of J'_{gk} within a criterion. Then, the information content (the content of one criterion that is uncorrelated with other criterions) of the criterion f_g is defined by L_g.

$$L_g = S_g \times C_g = S_g \sum_{g=1}^{m}(1 - \alpha_{ig}). \tag{2.46}$$

where w_g denotes the objective weight of criterion f_g, which can be expressed as

$$w_g = \frac{L_g}{\sum_{g=1}^{m} L_g}. \tag{2.47}$$

The weight matrix W is expressed as

$$W = [w_1\ w_2 \ldots\ w_m]^T. \tag{2.48}$$

Using $\boldsymbol{B}'$ to denote a new ideal characteristic vector, we can express a new attribute characteristic vector $\boldsymbol{\phi}'(\boldsymbol{x}_k)$ for the component x_k as

$$\boldsymbol{B}' = W\boldsymbol{B}, \tag{2.49}$$

and

$$\begin{aligned}\boldsymbol{\phi}'(\boldsymbol{x}_k) &= W\boldsymbol{\phi}(\boldsymbol{x}_k)\\ &= [w_1 J'_{1k}\ w_2 J'_{2k} \ldots\ w_m J'_{mk}]^T.\end{aligned} \tag{2.50}$$

Thus, the MCIM, which considers the correlation between criteria $I_k^{mc'}$, is defined in Eq. (2.51).

$$I_k^{mc'} = d\left(\boldsymbol{B}', \boldsymbol{\phi}'(x_k)\right) = \sqrt{\sum_{g=1}^{m} w_g^2 (1 - J'_{gk})^2}. \tag{2.51}$$

The γth component of the θth layer is defined by $x_{\gamma^\theta} (\theta, \gamma \in N)$. X_θ contains all components within layer θ, and $\boldsymbol{X}$ denotes the system and contains all the components within the system. Under the criterion f_g, $J_{g(\gamma^\theta|\theta)}$ represents the sensitivity of the x_{γ^θ}'s attribute to the X_θ's attribute, which is shown as Eq. (2.52).

$$J_{g(\gamma^\theta|\theta)} = \frac{\partial v_g(X_\theta)}{\partial v_g(x_{\gamma^\theta})}. \tag{2.52}$$

According to the PageRank algorithm [40], for a component $x_{\gamma^{\theta+1}}$ at layer $\theta + 1$, the sensitivity of x_{γ^θ}'s attribute to the $x_{\gamma^{\theta+1}}$'s attribute is defined by $J_{g(\gamma^\theta|\gamma^{\theta+1})}$ and calculated using Eq. (2.53).

$$J_{g(\gamma^\theta|\gamma^{\theta+1})} = \frac{\partial v_g(X_{\gamma^{\theta+1}})}{\partial v_g(x_{\gamma^\theta})} = \frac{J_{g(\gamma^\theta|\theta)}}{O_{\gamma^\theta}}. \tag{2.53}$$

The O_{γ^θ} is the number of chains out of the component x_{γ^θ}. $J_{g(\gamma^\theta|\theta+1)}$ represents the sensitivity of the x_{γ^θ}'s attribute to the $X_{\theta+1}$'s attribute under the gth criterion in multiple criteria, which is calculated by using Eq. (2.54).

$$J_{g(\gamma^\theta|\theta+1)} = \frac{\partial v_g(X_{\theta+1})}{\partial v_g(x_{\gamma^\theta})}. \tag{2.54}$$

Then, according to Eqs. (2.52) and (2.53), we have

$$\begin{aligned} J_{g(\gamma^\theta|\theta+1)} &= \frac{\partial v_g(X_{\theta+1})}{\partial v_g(x_{\gamma^\theta})} \\ &= \frac{\partial v_g(X_{\theta+1})}{\partial v_g(x_{\gamma^{\theta+1}})} \times \frac{\partial v_g(x_{\gamma^{\theta+1}})}{\partial v_g(x_{\gamma^\theta})} \\ &= J_{g(\gamma^{\theta+1}|\theta+1)} J_{g(\gamma^\theta|\gamma^{\theta+1})}. \end{aligned} \tag{2.55}$$

In a chain consisting of x_{γ^θ}, $x_{\gamma^{\theta+1}}, \ldots$, and $x_{\gamma^{\theta+\Delta}}$, $J_{g(\gamma^\theta|\gamma^{\theta+\Delta})}$ represents the sensitivity of the x_{γ^θ}'s attribute to the $x_{\gamma^{\theta+\Delta}}$'s attribute under the f_g criterion, which can express the indirect impact of upstream nodes on downstream nodes.

$$J_{g(\gamma^\theta|\gamma^{\theta+\Delta})} = \frac{\partial v_g(X_{\gamma^{\theta+\Delta}})}{\partial v_g(x_{\gamma^\theta})}. \tag{2.56}$$

Then, based on Eqs. (2.46) and (2.50), Eq. (2.56) can be expanded as follows.

$$
\begin{aligned}
J_{g(\gamma^{\theta}|\gamma^{\theta+\Delta})} &= \frac{\partial v_g(X_{\gamma^{\theta+\Delta}})}{\partial v_g(x_{\gamma^{\theta}})} = \frac{\partial v_g(X_{\gamma^{\theta+1}})}{\partial v_g(x_{\gamma^{\theta}})} \times \frac{\partial v_g(X_{\gamma^{\theta+2}})}{\partial v_g(x_{\gamma^{\theta+1}})} \times \cdots \times \frac{\partial v_g\left(X_{\gamma^{\theta+\Delta}}\right)}{\partial v_g\left(x_{\gamma^{\theta+\Delta-1}}\right)} \\
&= J_{g(\gamma^{\theta}|\gamma^{\theta+1})} \times J_{g(\gamma^{\theta+1}|\gamma^{\theta+2})} \times \cdots \times J_{g(\gamma^{\theta+\Delta-1}|\gamma^{\theta+\Delta})} \\
&= \sum_{\theta=1}^{\Delta-1} J_{g(\gamma^{\theta}|\gamma^{\theta+1})} \\
&= \frac{\sum_{\theta=1}^{\Delta-1} J_{g(\gamma^{\theta}|\theta)}}{\sum_{\theta=1}^{\Delta-1} O_{\gamma^{\theta}}}.
\end{aligned}
\tag{2.57}
$$

According to Eq. (2.57), in a series hierarchy, components in the downstream of the chain can choose favourable upstream components. In addition, for chains that start at $x_{\gamma^{\theta}}$ and end at $x_{\gamma^{\theta+\Delta}}$, the chain with the highest $J_{g(\gamma^{\theta}|\gamma^{\theta+\Delta})}$ is the most probable to magnify the hazard and influence the stable operation of $x_{\gamma^{\theta+\Delta}}$. For components $x_{\gamma^{\theta}}$ and $x_{\gamma^{\theta+\Delta}}$ that are working, choosing a chain with a smaller $J_{g(\gamma^{\theta}|\gamma^{\theta+\Delta})}$ value is beneficial for reducing cooperation risks.

According to Eqs. (2.52) and (2.57), we have

$$
\begin{aligned}
J_{g(\gamma^{\theta}|\theta+\Delta)} &= \frac{\partial v_g(X_{\theta+\Delta})}{\partial v_g(x_{\gamma^{\theta}})} \\
&= \frac{\partial v_g(X_{\theta+\Delta})}{\partial v_g(x_{\gamma^{\theta+\Delta}})} \times \frac{\partial v_g(x_{\gamma^{\theta+\Delta}})}{\partial v_g(x_{\gamma^{\theta}})}.
\end{aligned}
\tag{2.58}
$$

Equation (2.58) can help select upstream components that contribute to the stable operation of the downstream subsystem, thereby optimising the downstream subsystem.

2.1.2 Importance Measure of System Reliability Upgrade for Multi-state Consecutive k-out-of-n Systems

This section proposes a new importance measure of studying the reliability upgrade rate in multi-state consecutive k-out-of-n systems.

The state space of each component and system is $\{0, 1, 2, \ldots, M\}$, where 0 corresponds to the complete failure of the system or its component and M is the perfect functioning state of the system or component. The states are arranged in order from 0 to M. All components are statistically independent.

For a multi-state system, $\Phi(\boldsymbol{X}) \geq m$ represents that the system is working, where X represents the vector of states of the components $(x_1, x_2, \ldots, x_n)$ and x_i represents the state of component i, $x_i = 0, 1, 2, \ldots, M$. The system fails when $\Phi(\boldsymbol{X}) < m$. Hence, the Birnbaum importance measure of component i of the multi-state system is given by [46],

$$I_{i_m}^{\mathrm{B}} = \Pr\{\Phi(\boldsymbol{X}) \geq m|X_i \geq m\} - \Pr\{\Phi(\boldsymbol{X}) \geq m|X_i < m\}. \tag{2.59}$$

According to Eq. (2.59), the Birnbaum's importance measure for the consecutive k-out-of-n: F system in multi-state systems can be extended to Eq. (2.60).

$$\begin{aligned} I_{i_m}^{\mathrm{B}} &= \frac{\partial R(n)}{\partial P_{i_m}} = \mathrm{R}\,(n \mid X_i \geq m) - R(n|X_i < m) \\ &= \frac{R(i-1)R_{n-i}^{'} - R(n)}{1 - P_{i_m}}, \end{aligned} \tag{2.60}$$

where $P_{i_m} = \Pr(X_i \geq m)$.

According to Eq. (2.60), we have

$$\begin{aligned} R(n) &= P_{i_m} R\,(n \mid X_i \geq m) + \left(1 - P_{i_m}\right) R(n|X_i < m) \\ &= P_{i_m}\,[R\,(n \mid X_i \geq m) - R(n|X_i < m)] + R(n|X_i < m) \\ &= P_{i_m} I_{i_m}^{\mathrm{B}} + R(n|X_i < m). \end{aligned} \tag{2.61}$$

Improving the reliability of component i is equivalent to an upgrade of P_{i_m}, where $P_{i_m}^{\prime}$ is the reliability for the improved component i and $\overline{R}(n)$ is the system reliability after the improvement of component i. According to Eq. (2.61), the system reliability upgrade is given by

$$\begin{aligned} \overline{R}(n) - R(n) &= P_{i_m}^{'} I_i^{\mathrm{B}} + R\,(n \mid X_i < m) - P_{i_m} I_i^{\mathrm{B}} - R\,(nX_i < m) \\ &= (P_{i_m}^{'} - P_{i_m}) I_{i_m}^{\mathrm{B}}. \end{aligned} \tag{2.62}$$

However, for engineers, it is more meaningful to consider the variation of the system reliability upgrade rate with the improvement time of component i rather than the value of the system reliability upgrade. Therefore, when considering the average impact of improving component i per unit of time on system reliability upgrades, it is necessary to introduce the improvement rate (μ_i) of component i into Eq. (2.62), where μ_i represents the expected times that component i improves from P_{i_m} to $P_{i_m}^{'}$ in a unit of time. Hence, the rate of the system reliability upgrade based on the improvement of component i is

$$\mu_i\left(\overline{R}(n) - R(n)\right) = \mu_i (P_{i_m}^{'} - P_{i_m}) I_{i_m}^{\mathrm{B}}. \tag{2.63}$$

Definition 2.1 The rate of system reliability upgrade caused by the improvement of component i in a multi-state consecutive k-out-of-n: F system based on the system performance level m can be defined by the increased potential importance (IPI) of component i, I_i^{IP}, as shown below.

$$
\begin{aligned}
I_i^{IP} &= \mu_i \left(P_{i_m}^{'} - P_{i_m}\right) I_i^{\mathrm{B}} \\
&= \mu_i (P_{i_m}^{'} - P_{i_m}) \frac{R(i-1)R_{n-i}^{'} - R(n)}{1 - P_{i_m}}.
\end{aligned} \tag{2.64}
$$

The physical meaning of I_i^{IP} is the rate of the system reliability upgrade caused by the improvement of component i. For a multi-state consecutive k-out-of-n: F system consisting of n components, we should choose the most important component which represents the maximisation of the rate of the system reliability upgrade based on I_i^{IP}.

Based on Definition 2.1, we have $I_i^{IP} = \mu_i \left(P_{i_m}' - P_{i_m}\right) I_i^{\mathrm{B}}$. According to some different improvement actions, we can obtain the following three scenarios.

(1) If the improvement action is to upgrade the component reliability by the same amount, $P_{i_m}' - P_{i_m} = A$, A is a constant, for all component i, then we have $I_i^{IP} = A\mu_i I_i^B$. If $\mu_1 = \mu_2 = \cdots = \mu_n = \mu$, then $I_i^{IP} = A\mu I_i^B$.
(2) If component i is improved to its perfect functioning, then $P_{i_m}^{'} = 1$. we can obtain

$$
\begin{aligned}
I_i^{IP} &= \mu_i \left(1 - P_{i_m}\right) I_i^B \\
&= \mu_i \left(1 - P_{i_m}\right) \frac{R(i-1)R_{n-i}^{'} - R(n)}{1 - P_{i_m}} \\
&= \mu_i [R(i-1)R_{n-i}^{'} - R(n)].
\end{aligned} \tag{2.65}
$$

(3) If parallel redundancy is used to represent the improvement of component reliability, then $P_{i_m}^{'} = 1 - (i - P_{i_m})^2$. Hence, we can obtain

$$
\begin{aligned}
I_i^{IP} &= \mu_i \left(P_{i_m}^{'} - P_{i_m}\right) I_i^{\mathrm{B}} \\
&= \mu_i \left(1 - (i - P_{i_m})^2 - P_{i_m}\right) I_i^{\mathrm{B}} \\
&= \mu_i P_{i_m} [R(i-1)R_{n-i}^{'} - R(n)].
\end{aligned} \tag{2.66}
$$

Theorem 2.4 *The largest improvement in system reliability per unit of time is to improve the component with the maximum value I_i^{IP} $(1 \le i \le n)$ for a multi-state consecutive k-out-of-n: F system.*

Proof Immediately from Eqs. (2.63) and (2.64).

Using $R_L(j)$, $R_L'(j)$, and $R_L(n)$ to represent $R(j)$, $R'(j)$, $R(n)$ of the linear system, we can obtain Theorem 2.4. □

Theorem 2.5 *For a multi-state linear consecutive k-out-of-n: F system, we can obtain that*

$$I_i^{IP} = \mu_i(P'_{i_m} - P_{i_m})\frac{R_L(i-1)R'_L(n-i) - R_L(n)}{1 - P_{i_m}}. \tag{2.67}$$

Proof According to Eq. (2.60), we have

$$\begin{aligned} I_{i_m}^{\mathrm{B}} &= \frac{\partial R(n)}{\partial P_{i_m}} \\ &= R(n|x_i \geq m) - R\,(n|x_i < m) \\ &= \frac{R(i-1)R'_{(n-i)} - R(n)}{1 - P_{i_m}}. \end{aligned} \tag{2.68}$$

For a multi-state linear consecutive k-out-of-n: F system, we can obtain $R_L(i-1) = R(i-1)$, $R'_L(n-i) = R'(n-i)$ and $R_L(n) = R(n)$. According to Eq. (2.64), we can obtain that

$$I_i^{\mathrm{IP}} = \mu_i(P'_{i_m} - P_{i_m})\frac{R_L(i-1)R'_L(n-i) - R_L(n)}{1 - P_{i_m}}. \tag{2.69}$$

□

Corollary 2.1 *If all components are independent and identically distributed (i.i.d), and $P_{1,m} = P_{2,m} = \cdots = P_{n,m} = P$, $P'_{1,m} = P'_{2,m} = \cdots = P'_{n,m} = P'$, then we can obtain that $I_i^{\mathrm{IP}} = \mu_i\left(P' - P\right)\frac{R_L(i-1)R_L(n-i) - R_L(n)}{1-P}$ for a multi-state linear consecutive k-out-of-n: F system.*

Proof If all components are independent and identically distributed, then we can obtain that

$$R'_L(n-i) = R_L\left(P_{i+1,m}, \ldots, P_{N,M}\right) = R_L\left(\underbrace{P, \ldots, P}_{N-1}\right) = R_L(n-i). \tag{2.70}$$

According to Theorem 2.5, we have that

$$I_i^{IP} = \mu_i\left(P' - P\right)\frac{R_L(i-1)R_L(n-i) - R_L(n)}{1 - P}. \tag{2.71}$$

Using $R_C(j)$, $R'_C(j)$, and $R_C(n)$ to represent the $R(j)$, $R'(j)$, and $R(n)$ of the circular system, respectively, we can obtain Theorem 2.6.

Theorem 2.6 *For a multi-state circular consecutive k-out-of-n: F system, we can obtain that $I_i^{\mathrm{IP}} = \mu_i\left(P'_{i_m} - P_{i_m}\right)\frac{R_L(i+1,i-1) - R_C(n)}{1-P_{i_m}}$, where $R_L(i+1, i-1) = R_L\left(P_{i+1,m}, \ldots, P_{N,M}, P_{1,m}, \ldots, P_{i-1,m}\right)$.*

Proof According to Eq. (2.60), we have

$$\begin{aligned} I_i^{\mathrm{B}} &= \frac{\partial R(n)}{\partial P_{i_m}} \\ &= R\left(n|x_i \geq m\right) - R\left(n|x_i < m\right) \\ &= \frac{R(i-1)R'(n-i) - R(n)}{1 - P_{i_m}}. \end{aligned} \tag{2.72}$$

For a circular multi-state consecutive k-out-of-n: F system, we can obtain that $R(n) = R_C(n)$ and

$$\begin{aligned} R(i-1)R'(n-i) &= R_C(i-1)R'_C(n-i) \\ &= R_C\left(P_{1,m}, \ldots, P_{i-1,m}, x_i \geq m, P_{i+1,m}, \ldots, P_{N,M}\right) \\ &= R_L\left(P_{i+1,m}, \ldots, P_{N,M}, P_{1,m}, \ldots, P_{i-1,m}\right) \\ &= R_L(i+1, i-1). \end{aligned} \tag{2.73}$$

Hence, according to Eq. (2.64), we can obtain

$$I_i^{IP} = \mu_i\left(P'_{i_m} - P_{i_m}\right)\frac{R_L(i+1, i-1) - R_C(n)}{1 - P_{i_m}}. \tag{2.74}$$

Corollary 2.2 *If all components are i.i.d, and $P_{1,m} = P_{2,m} = \cdots = P_{n,m} = P$, $P'_{1,m} = P'_{2,m} = \cdots = P'_{n,m} = P'$, then we can obtain that $I_i^{IP} = \mu_i(P' - P)$ $\frac{R_L(n-1) - R_C(n)}{1-P}$ for a multi-state circular consecutive k-out-of-n: F system.*

Proof If all components are i.i.d, then we can obtain that

$$\begin{aligned} R_L(i+1, i-1) &= R_L\left(P_{i+1,m}, \ldots, P_{N,M}, P_{1,m}, \ldots, P_{i-1,m}\right) \\ &= R_L\left(\underbrace{P, \ldots, P}_{n-1}\right) \\ &= R_L(n-1). \end{aligned} \tag{2.75}$$

According to Theorem 2.6, we have that $I_i^{IP} = \mu_i(P' - P)\frac{R_L(n-1) - R_C(n)}{1-P}$. □

Corollary 2.3 *Suppose all components are i.i.d, and $P_{1,m} = P_{2,m} = \cdots = P_{n,m} = P$, $P'_{1,m} = P'_{2,m} = \cdots = P'_{n,m} = P'$, If $\mu_i \geq \mu_j$, then $I_i^{IP} \geq I_j^{IP}$ for a multi-state circular consecutive k-out-of-n: F system.*

Proof The result can be established by mimicking the proof in Corollary 2.2. □

Theorem 2.7 *In the multi-state linear consecutive k-out-of-n: F system based on the system performance level m, if all components are i.i.d, and $P_{1,m} = P_{2,m} = \cdots = P_{n,m} = P$, $P'_{1,m} = P'_{2,m} = \cdots = P'_{n,m} = P'$, then we have*

- *if* $\mu_i \le \mu_{i+1}$, *then* $I_i^{IP} \le I_{i+1}^{IP}$ *for* $n - k_m + 1 \le i < k_m$.
- *when* $\frac{n}{2} < k_m < n$, *if* $\mu_i \le \mu_{i+1}$, *then* $I_i^{IP} < I_{i+1}^{IP}$ *for* $i + 1 \le n - k_m + 1$.
- *when* $2 < k_m \le \frac{n}{2}$, *if* $\mu_i \ge \mu_{i+1}$, *then* $I_i^{IP} > I_{i+1}^{IP}$ *for* $i > n - k_m$.
- *when* $2 < k_m \le \frac{n}{2}$, *if* $\mu_i \le \mu_{i+1}$, *then* $I_i^{IP} < I_{i+1}^{IP}$ *for* $i < k_m$.

Proof According to Corollary 2.1, we have

$$I_i^{IP} = \mu_i (P' - P) \frac{R_L(i-1) R_L(n-i) - R_L(n)}{1 - P} \tag{2.76}$$

and

$$I_{i+1}^{IP} = \mu_i (P' - P) \frac{R_L(i) R_L(n-i-1) - R_L(n)}{1 - P}. \tag{2.77}$$

- Since $i < k_m$, we can obtain $R_L(i) = 1$ and $R_L(i-1) = 1$. Similarly, since $n - k_m + 1 \le i$, we have $R_L(n-i) = 1$ and $R_L(n-i-1) = 1$. Then we can obtain $I_i^{IP} = \mu_i (P' - P) \frac{1 - R_L(n)}{1-P}$, $I_{i+1}^{IP} = \mu_{i+1} (P' - P) \frac{1 - R_L(n)}{1-P}$. Hence, if $\mu_i \le \mu_{i+1}$, then $I_i^{IP} \le I_{i+1}^{IP}$.
- When $\frac{n}{2} < k_m < n$, we have $n - k_m < k_m$. Since $i + 1 \le n - k_m + 1$, we can obtain $i + 1 \le n - k_m + 1 \le k_m$. Then we have $R_L(i) = 1$ and $R_L(i-1) = 1$. Hence, we can obtain $I_i^{IP} = \mu_i (P' - P) \frac{R_L(n-i) - R_L(n)}{1-P}$ and $I_{i+1}^{IP} = \mu_{i+1} (P' - P) \frac{R_L(n-i-1) - R_L(n)}{1-P}$. Since $R_L(n-i) < R_L(n-i-1)$, when $\mu_i \le \mu_{i+1}$, we can obtain $I_i^{IP} < I_{i+1}^{IP}$.
- When $2 < k_m \le \frac{n}{2}$, we have $n - k_m \ge k_m$. Since $n - k_m > i$, we can obtain $k_m \le n - k_m < i$. Then we have $R_L(n-i) = 1$ and $R_L(n-i-1) = 1$. Hence, we can obtain $I_i^{IP} = \mu_i (P' - P) \frac{R_L(i-1) - R_L(n)}{1-P}$ and $I_{i+1}^{IP} = \mu_{i+1} (P' - P) \frac{R_L(i) - R_L(n)}{1-P}$. Since $R_L(i) < R_L(i-1)$, when $\mu_i \ge \mu_{i+1}$, we can obtain $I_i^{IP} > I_{i+1}^{IP}$.
- When $2 < k_m \le \frac{n}{2}$, we have $n - k_m \ge k_m$. Since $i < k_m$, we can obtain $i < k_m \le n - k_m$. Then we have $R_L(i) = 1$ and $R_L(i-1) = 1$. Hence, we can obtain that $I_i^{IP} = \mu_i (P' - P) \frac{R_L(n-i) - R_L(n)}{1-P}$ and $I_{i+1}^{IP} = \mu_{i+1} (P' - P) \frac{R_L(n-i-1) - R_L(n)}{1-P}$. Since $R_L(n-i) < R_L(n-i-1)$, when $\mu_i \le \mu_{i+1}$, we can obtain $I_i^{IP} < I_{i+1}^{IP}$.

This completes the proof of the theorem. □

From Corollary 2.3, the IP order of all components in the circular consecutive k-out-of-n: F system can be obtained. From Theorem 2.7, the IP order of the adjacent components in a linear consecutive k-out-of-n: F system can be obtained.

For decreasing or increasing multi-state consecutive k-out-of-n: F systems, $I_{i_m}^{IP}$ means the increase potential importance of component i based on system performance level m. μ_{i_m} means the expected time for component i to improve from P_{i_m} to P'_{i_m} in a unit of time based on system performance level m, $R_m(j)$ means the reliability of multi-state consecutive k-out-of-j: F subsystem consisting of components $1, 2, \ldots, j$ based on system performance level m. $R'_m(j)$ means the reliability of the multi-state consecutive k-out-of-j: F subsystem composed of components

$(n-j+1), (n-j+2), \ldots, (n-1), n$, based on system performance level m. We can obtain the following theorems.

Theorem 2.8 *For a decreasing multi-state consecutive k-out-of-n: F system based on two different system performance level r, s, $r < s$, if $P'_{i,r} - P_{i,r} = P'_{i,s} - P_{i,s} = \Delta$ and $\mu_{i,r} = \mu_{i,s} = \mu$, then we can obtain that $I^{IP}_{i,s} \leq I^{IP}_{i,r}$.*

Proof According to Eq. (2.64), we have $I^{IP}_{i,r} = \mu\Delta\frac{R_r(i-1)R'_r(n-i)-R(n)}{1-P_{i,r}}$ and $I^{IP}_{i,s} = \mu\Delta\frac{R_s(i-1)R'_s(n-i)-R(n)}{1-P_{i,s}}$. Since $r < s$, we obtain $P_{i,r} \geq P_{i,s} \Rightarrow 1 - P_{i,r} \leq 1 - P_{i,s}$. For a decreasing multi-state consecutive k-out-of-n: F system, we have $k_r \geq k_s$, so, $R_r(i-1) \geq R_s(i-1)$, $R'_r(n-i) \geq R'_s(n-i)$. Then we can obtain $I^{IP}_{i,s} \leq I^{IP}_{i,r}$. □

Corollary 2.4 *For a decreasing (increasing) multi-state consecutive k-out-of-n: F system based on two different system performance level r, s, $r < s$, if component i is improved to the perfect functioning and $\mu_{i,r} = \mu_{i,s} = \mu$, then $I^{IP}_{i,s} \leq I^{IP}_{i,r}$ ($I^{IP}_{i,s} \geq I^{IP}_{i,r}$).*

Proof According to Eq. (2.65), we have $I^{IP}_{i,r} = \mu[R_r(i-1)R'_r(n-i) - R(n)]$ and $I^{IP}_{i,s} = \mu[R_s(i-1)R'_s(n-i) - R(n)]$. The rest proof is similar to Theorem 2.8. □

The IP of the target component is the rate of system reliability upgrade caused by the improvement of the target component. The optimal improvement in system reliability per unit of time is determined by the maximum IP of all components. Even if one component generates a larger Birnbaum importance measure than another, its IP may still be low. For engineers, IP is a good measure to determine which component i is the most responsible for the system reliability upgrade rate.

2.2 Importance Measures for System Reconfiguration

Usually, importance measures do not consider the changes that may occur in the system structure as the reliability of specific component improves. However, if the reliability of components changes, the optimal system configuration may also change, and the importance measure of the corresponding components also changes according to the selected structure.

2.2.1 *Introduction*

For a heterogeneous system [13–16, 32, 35], the reliability of the system can be improved by optimizing the ordering or the weighting of the components for maintenance. When the parameters of system components are changed, the system can easily be re-arranged to achieve the maximum reliability. Therefore, the study of the possible effects of component reliability improvement must take into account the optimal system reconfiguration after such improvement. The optimal configuration problem was studied by Malon [29, 30] and Tong [39]. Zuo and Kuo [48] summarised the existing progress on invariant optimal designs for continuous k-out-of-n and identified invariant optimal designs for such systems. Levitin and Lisnianski [23] studied reliability optimisation of weighted voting systems. The articles by Levitin [20, 21]; Levitin et al. [24] studied linear continuously connected systems, multi-state sliding window systems, and the optimal ordering of components in heterogeneous standby systems. For a linear and circular continuous k-out-of-n: F system, Yun et al. [43] analysed the system configuration parameter k, and the effect of various cost parameters on the optimal n. Boddu and Xing [2, 3] considered the redundancy allocation problem associated with k-out-of-n systems, and in order to obtain the optimal design configuration of the system, proposed a penalty-guided genetic algorithm based optimisation solution. Steffen et al. [38] analysed the reliability of actuating elements in series and parallel systems and gave the optimal configuration for satisfying the pre-defined requirements.

Therefore importance measures are also used to solve the component assignment problem for continuous k-out-of-n systems. Kuo et al. [19] gave heuristics for the optimal design of general linear continuous-k-out-of-n systems. Hwang et al. [17], Chang et al. [7], and Chang et al. [6] discussed the application of the Birnbaum importance to continuous-k-out-of-n: F systems and continuous-k-out-of-n: G systems. Lin and Kuo [25] studied the reliability of the system allocation and the Birnbaum importance, which can be used to examine whether a system has any invariant optimal allocation. Yao et al. [42] proposed five heuristics based on the Birnbaum importance and analysed their corresponding properties. Zhu et al. [44] studied the relationship between the Birnbaum importance and the reliability of common components and summarised the Birnbaum importance methods for linear continuous k-out-of-n systems.

2.2.2 *Importance Measure Analysis for Reconfigurable Systems*

The goal of system structure optimisation is to allocate components to n positions to maximise the reliability of the system.

Definition 2.2 The optimal system configuration is a permutation $\tau^* \in \mathrm{S}$ such that

$$\phi(\tau^*, p) = \max_{\tau \in S} \phi(\tau, p), \tag{2.78}$$

where $\phi(\tau, p)$ is the system reliability corresponding to the permutation t.

When the reliability of component located in a specific location is changed from $p_{\tau(i)}$ to $p^*_{\tau(i)}$, and the reliability of all other components remains unchanged, the system reliability becomes

$$\begin{aligned}\phi\left(\tau^*, p\right) = & p^*_{\tau(i)} p_{\tau(i)} R\left(1_i, 1_j, p\right) \\ & + p^*_{\tau(i)}\left(1 - p_{\tau(j)}\right) R\left(1_i, 0_j, p\right) \\ & + \left(1 - p^*_{\tau(i)}\right)\left(1 - p_{\tau(j)}\right) R\left(0_i, 0_j, p\right) \\ & + \left(1 - p^*_{\tau(i)}\right) p_{\tau(j)} R\left(0_i, 1_j, p\right). \end{aligned} \tag{2.79}$$

After the reliability $p_{\tau(i)}$ changes, the arrangement may become suboptimal. The exchange of components initially resident in positions $\tau(i)$ and $\tau(j)$ makes system reliability become

$$\begin{aligned}\phi(\tau^*_{ij}, p) = & p^*_{\tau(i)} p_{\tau(j)} R(1_i, 1_j, p) + p_{\tau(j)}(1 - p^*_{\tau(i)}) R(1_i, 0_j, p) \\ & + (1 - p^*_{\tau(i)})(1 - p_{\tau(j)}) R(0_i, 0_j, p) + (1 - p_{\tau(j)}) p^*_{\tau(i)} R(0_i, 1_j, p). \end{aligned} \tag{2.80}$$

The variation of the system reliability after this interchange is

$$\begin{aligned} & \phi(\tau^*, p) - \phi(\tau^*_{ij}, p) \\ & = p^*_{\tau(i)}\left(1 - p_{\tau(j)}\right) R\left(1_i, 0_j, p\right) + \left(1 - p^*_{\tau(i)}\right) p_{\tau(j)} R(0_i, 1_j, p) \\ & \quad - p_{\tau(j)}\left(1 - p^*_{\tau(i)}\right) R\left(1_i, 0_j, p\right) - \left(1 - p_{\tau(j)}\right) p^*_{\tau(i)} R(0_i, 1_j, p) \\ & = p^*_{\tau(i)} - p_{\tau(j)} R\left(1_i, 0_j, p\right) - \left(p^*_{\tau(i)} - p_{\tau(j)}\right) R(0_i, 1_j, p) \\ & = \left(p^*_{\tau(i)} - p_{\tau(j)}\right)\left(R\left(1_i, 0_j, p\right) - R(0_i, 1_j, p)\right). \end{aligned} \tag{2.81}$$

If $\phi(\tau^*, p) - \phi(\tau^*_{ij}, p) < 0$, then one obtains better permutation by interchanging the components in positions $\tau(i)$ and $\tau(j)$.

In a series system, $\phi(\tau, \boldsymbol{p}) = \sum_{j=1}^{n} p_{\tau(j)}$ and in a parallel system, $\phi(\tau, \boldsymbol{p}) = 1 - \sum_{j=1}^{n}(1 - p_{\tau(j)})$. So for any permutation t, the corresponding system reliability is the same.

In a k-out-of-n system, $R(1_i, 0_j, \boldsymbol{p})$ and $R(0_i, 1_j, \boldsymbol{p})$ represents the reliability of a (k-1)-out-of-($n - 2$) system consisting of the $n - 2$ components in positions $\tau(x)$, $1 \le x \le n$ and $x \ne i, j$. So $R(1_i, 0_j, \boldsymbol{p}) = R(0_i, 1_j, \boldsymbol{p})$. Then, in a k-out-of-n system, for the optimal configuration, the permutation remains optimal when the reliability of the component at a specific location changes.

In a consecutive k-out-of-n: G systems, assume that $\tau(i) < \tau(j) \le k, n \ge 2k$. $R(1_i, 0_j, \boldsymbol{p})$ represents the reliability of a linear continuous k-out of (n-j): G system composed of $(n - j)$ components at positions $\tau(x)$, $j + 1 \le x \le n$. $R(0_i, 1_j, \boldsymbol{p})$ represents the reliability of a linear consecutive k-out-of-($n - j + 1$): G system composed of ($n - j + 1$) components at positions $\tau(x)$, $j \le x \le n$. Obviously, $R(1_i, 0_j, \boldsymbol{p}) \le R(0_i, 1_j, \boldsymbol{p})$. If $p^*_{\tau(i)} < p_{\tau(j)}$, then $\phi(\tau^*, \boldsymbol{p}) \ge \phi(\tau^*_{ij}, \boldsymbol{p})$. If $p^*_{\tau(i)} > p_{\tau(j)}$, then $\phi(\tau^*, \boldsymbol{p}) \le \phi(\tau^*_{ij}, \boldsymbol{p})$. Therefore, if $p^*_{\tau(i)} > p_{\tau(j)}$, then a better permutation can be obtained by swapping positions $\tau(i)$ and $\tau(j)$.

In a consecutive k-out-of-n: F system, we assume that $\tau(i) < \tau(j) \le k, n \ge 2k$. $Q(1_i, 0_j, \boldsymbol{p})$ represents the reliability of a linear continuous k-out-of-($n - j + 1$):

F system composed of $(n - j + 1)$ components at positions $\tau(x), j \leq x \leq n$. $Q(0_i, 1_j, \boldsymbol{p})$ represents the reliability of a linear consecutive k-out-of-$(n - j)$: F system composed of $(n - j)$ components at positions $\tau(x), j + 1 \leq x \leq n$. Obviously, $Q(1_i, 0_j, \boldsymbol{p}) \geq Q(0_i, 1_j, \boldsymbol{p})$. According to Eq. (2.81), if $p^*_{\tau(i)} < p_{\tau(j)}$, then $\phi(\tau^*, \boldsymbol{p}) \geq \phi(\tau^*_{ij}, \boldsymbol{p})$. If $p^*_{\tau(i)} > p_{\tau(j)}$, then $\phi(\tau^*, \boldsymbol{p}) \geq \phi(\tau^*_{ij}, \boldsymbol{p})$. Therefore, if $p^*_{\tau(i)} > p_{\tau(j)}$, then a better permutation can be obtained by swapping positions $\tau(i)$ and $\tau(j)$.

If the reliability of components changes, the optimal configuration of the system may also change. Then, the importance of the corresponding components will change as the optimal structure of the system changes. The importance measure indicates which component's reliability improvement is most promising if the system structure can be changed.

In optimal permutation τ^*, if the reliability of a component is changed from p_i to $p_i + \Delta$ and the reliability of other components remains unchanged. then the system reliability is

$$\begin{aligned} R\left((p_i + \Delta)_i, p\right) &= (p_i + \Delta)\, R\,(1_i, p) + (1 - p_i - \Delta)\, R\,(0_i, p) \\ &= p_i R\,(1_i, p) + (1 - p_i)\, R\,(0_i, p) + \Delta \times (R\,(1_i, p) - R\,(0_i, p)) \\ &= R\,(p) + \Delta I_i^B. \end{aligned} \tag{2.82}$$

From Eq. (2.82), it can be seen that when the reliability of a specific component increases Δ, the change in system reliability is $\Delta \mathrm{I}_i^B$. However, after the reliability of specific components changes, the original permutation may no longer be optimal. Changes in component sequences can also affect the Birnbaum importance measure of specific components.

The approximation of the importance measure for component i is given by

$$I_i(\Delta, p_i) = \frac{R^{\mathrm{opt}}(p_i + \Delta, \mathrm{p}) - R^{\mathrm{opt}}(p_i, p)}{\Delta}, \tag{2.83}$$

where $R^{\mathrm{opt}}(p_i, p)$ and $R^{\mathrm{opt}}(p_i + \Delta, p)$ are system reliabilities obtained for the optimal configurations, corresponding to component reliability vectors (p_i, p) and $(p_i + \Delta, p)$, respectively. Δ is a finite variation of the component reliability. Note that $I_i = \lim_{\Delta \to 0} I_i(\Delta, p_i)$.

The optimal configuration was obtained using the genetic algorithm. We can find the description of the algorithm and the solution to the optimal permutation problem in Levitin [21].

We have demonstrated the following attributes of the importance measures.

Changes in component reliability may affect the optimal structure of the system. The reliability of the optimal reconfigurable system monotonically (but not smoothly) increases with the reliability of any component.

The reliability importance measure of components in an optimal reconfigurable system may vary non-monotonically with respect of the reliability of that component.

The ranking of components based on their reliability importance may vary with changes in component reliability.

2.2.3 Importance Measures for System Reconfiguration in Linear Consecutive-k-out-of-n Systems

The mean absolute deviation (MAD) was used by Ramirez-Marquez and Coit [33, 34] to evaluate the effect of all states of a component on system reliability. The MAD of component i is defined by

$$I_i^{mad}(t) = \sum_l \{\Pr(X_i = 1) \,|\, \Pr(\Phi(l_i, \boldsymbol{X}) \geq d) - \Pr(\Phi(\boldsymbol{X}) \geq d)\}, \tag{2.84}$$

where d represents the system constant demand.

For binary systems, system constant demand d represents the working state of the system. Given that

$$\begin{aligned}\Pr\{\Phi(\boldsymbol{X}(t)) = 1\} = \Pr(X_i(t) = 1)\Pr\{\Phi(\boldsymbol{X}(t) = 1|X_i(t) = 1\}\\ + \Pr(X_i(t) = 0)\Pr\{\Phi(\boldsymbol{X}(t)) = 1|X_i(t) = 0\},\end{aligned} \tag{2.85}$$

we can obtain the following equation:

$$\begin{aligned}I_i^{mad}(t) &= \Pr(X_i(t) = 1)\{\Pr\{\Phi(\boldsymbol{X}(t)) = 1|X_i(t) = 1\} - \Pr\{\Phi(\boldsymbol{X}(t)) = 1\}\}\\ &\quad + \Pr(X_i(t) = 0)\{\Pr\{\Phi(\boldsymbol{X}(t)) = 1\} - \Pr\{\Phi(\boldsymbol{X}(t)) = 1|X_i(t) = 0\}\}\\ &= 2\Pr(X_i(t) = 1)\Pr(X_i(t) = 0)\{\Pr\{\Phi(\boldsymbol{X}(t)) = 1|X_i(t) = 1\}\\ &\quad - \Pr\{\Phi(\boldsymbol{X}(t)) = 1|X_i(t) = 0\}\}\\ &= 2\Pr(X_i(t) = 1)\Pr(X_i(t) = 0)\, I_i^{\mathrm{B}}(t).\end{aligned} \tag{2.86}$$

Subsequently, the MAD of component i for a binary system is

$$I_i^{mad}(t) = 2\Pr(X_i(t) = 1)\Pr(X_i(t) = 0)\, I_i^{\mathrm{B}}(t). \tag{2.87}$$

Papastavridis [32] presented the Birnbaum importance for the consecutive-k-out-of-n: F system as follows.

$$\begin{aligned}I_i^{\mathrm{B}}(t) &= \frac{\partial R^t(n)}{\partial P_{i_1}(t)}\\ &= R^t(n|X_i(t) = 1) - R^t(n|X_i(t) = 0)\\ &= \frac{R^t(i-1)R'^t(n-i) - R^t(n)}{1 - P_{i_1}(t)},\end{aligned} \tag{2.88}$$

where $R^t(j)$ is the reliability of the consecutive-k-out-of-j: F subsystem consisting of components $1, 2, \ldots, j$ at time t, and $R'^t(j)$ is the reliability of the consecutive-k-out-of-j: F subsystem consisting of components $(n-j+1), (n-j+2), \ldots, (n-1), n$.

According to Eq. (2.88), we can obtain the integrated importance measure for the consecutive-k-out-of-n: F system.

$$\begin{aligned} I_i^{im}(t) &= \Pr(X_i(t)=1)\,\lambda_i(t) I_i^{\mathrm{B}}(t) \\ &= \lambda_i(t) P_{i_1}(t) \frac{R^t(i-1)R'^t(n-i) - R^t(n)}{1 - P_{i_1}(t)}. \end{aligned} \tag{2.89}$$

According to Eq. (2.88), we can obtain the MAD for the consecutive-k-out-of-n: F system.

$$\begin{aligned} I_i^{mad}(t) &= 2\Pr(X_i(t)=1)\Pr(X_i(t)=0)\, I_i^{\mathrm{B}}(t) \\ &= 2P_{i_1}(t)\left[R^t(i-1)R'^t(n-i) - R^t(n)\right]. \end{aligned} \tag{2.90}$$

When the reliability of components changes, the optimal system structure may also change. Subsequently, the importance of the corresponding components changes as the optimal system structure changes. Assuming that $R^t_{opt}(i-1)$, $R'^t{}_{opt}(n-i)$, and $R^t_{opt}(n)$ are the reliabilities of the system or subsystem obtained for the optimal configurations at time t, then we can obtain the following definitions.

Definition 2.3 For the optimal structure of a consecutive-k-out-of-n: F system, the Birnbaum importance measure of component i is

$$I_i^{\mathrm{B}}(t) = \frac{R^t_{opt}(i-1)R'^t{}_{opt}(n-i) - R^t_{opt}(n)}{1 - P_{i_1}(t)}. \tag{2.91}$$

Definition 2.4 For the optimal structure of a consecutive-k-out-of-n: F system, the integrated importance measure of component i is

$$I_i^{im}(t) = P_{i_1}(t)\lambda_i(t)\frac{R^t_{opt}(i-1)R'^t{}_{opt}(n-i) - R^t_{opt}(n)}{1 - P_{i_1}(t)}. \tag{2.92}$$

Definition 2.5 For the optimal structure of a consecutive-k-out-of-n: F system, the MAD of component i is

$$I_i^{mad}(t) = 2P_{i_1}(t)\left[R^t(i-1)R'^t(n-i) - R^t(n)\right]. \tag{2.93}$$

In practice, the reliability of the components in a consecutive-k-out-of-n system (such as oil pipelines and street lighting systems) may be the same. In this section, the characteristics of the importance measure of the optimal structure when all components are independently and identically distributed are discussed.

Proposition 2.1 *For the optimal structure of a linear consecutive-k-out-of-n: F system, when all components are independently identically distributed, and*

$$P_{1_1}(t) = P_{2_1}(t) = \cdots = P_{n_1}(t) = P(t), \tag{2.94}$$

then

$$\begin{aligned} I_i^B(t) &= \frac{R^t_{opt}(i-1)R^t_{opt}(n-i) - R^t_{opt}(n)}{1-P(t)}, \\ I_i^{im}(t) &= P(t)\lambda_i(t)\frac{R^t_{opt}(i-1)R^t_{opt}(n-i) - R^t_{opt}(n)}{1-P(t)}, \\ I_i^{mad}(t) &= 2P(t)\left[R^t_{opt}(i-1)R^t_{opt}(n-i) - R^t_{opt}(n)\right]. \end{aligned} \tag{2.95}$$

Proof When all components are independently identically distributed,

$$\begin{aligned} R'^t(n-i) &= R^t(P_{i+1,K}(t), \ldots, P_{n,K}(t)) = R^t(\underbrace{P(t), \ldots, P(t)}_{n-i}) \\ &= R^t(n-i). \end{aligned} \tag{2.96}$$

According to Definitions 2.3, 2.4, and 2.5, we have

$$\begin{aligned} I_i^{\mathrm{B}}(t) &= \frac{R^t_{opt}(i-1)R^t_{opt}(n-i) - R^t_{opt}(n)}{1-P(t)}, \\ I_i^{im}(t) &= P(t)\lambda_i(t)\frac{R^t_{opt}(i-1)R^t_{opt}(n-i) - R^t_{opt}(n)}{1-P(t)}, \\ I_i^{mad}(t) &= 2P(t)\left[R^t_{opt}(i-1)R^t_{opt}(n-i) - R^t_{opt}(n)\right], \end{aligned} \tag{2.97}$$

respectively. □

According to Proposition 2.1, when all components are independently identically distributed, and $P_{1_1}(t) = P_{2_1}(t) = \cdots = P_{n1}(t) = P(t)$, then the Birnbaum importance measure and the MAD of all components are the same.

Proposition 2.2 *For the optimal structure of a linear consecutive-k-out-of-n: F system, all components are independently identically distributed, and* $P_{1_1}(t) = P_{2_1}(t) = \cdots = P_{n_1}(t) = P(t)$.

- *when* $\lambda_i(t) \le \lambda_{i+1}(t)$*, then* $I_i^{im}(t) \le I_{i+1}^{im}(t)$ *for* $n-k+1 \le i < k$.
- *when* $\frac{n}{2} < k < n$*, if* $\lambda_i(t) \le \lambda_{i+1}(t)$*, then* $I_i^{im}(t) \le I_{i+1}^{im}(t)$ *for* $i+1 \le n-k+1$.
- *when* $2 < k \le \frac{n}{2}$*, if* $\lambda_i(t) \ge \lambda_{i+1}(t)$*, then* $I_i^{im}(t) \ge I_{i+1}^{im}(t)$ *for* $i > n-k$.
- *when* $2 < k \le \frac{n}{2}$*, if* $\lambda_i(t) \le \lambda_{i+1}(t)$*, then* $I_i^{im}(t) \le I_{i+1}^{im}(t)$ *for* $i < k$.

Proof According to Proposition 2.1,

$$I_i^{im}(t) = P(t)\lambda_i(t)\frac{R_{opt}^t(i-1)R_{opt}^t(n-i) - R_{opt}^t(n)}{1 - P(t)} \tag{2.98}$$

and

$$I_{i+1}^{im}(t) = P(t)\lambda_{i+1}(t)\frac{R_{opt}^t(i-1)R_{opt}^t(n-i) - R_{opt}^t(n)}{1 - P(t)}. \tag{2.99}$$

Given that $i < k$, $R_{opt}^t(i) = 1$ and $R_{opt}^t(i-1) = 1$. Similarly, if $n-k+1 \leq i$, we can obtain $R_{opt}^t(n-i) = 1$ and $R_{opt}^t(n-i-1) = 1$. Then the component integrated importance measure can be expressed as $I_i^{im}(t) = P(t)\lambda_i(t)\frac{1-R_{opt}^t(n)}{1-P(t)}$, and $I_{i+1}^{im}(t) = P(t)\lambda_{i+1}(t)\frac{1-R_{opt}^t(n)}{1-P(t)}$. Hence, if $\lambda_i(t) \leq \lambda_{i+1}(t)$, then $I_i^{im}(t) \leq I_{i+1}^{im}(t)$.

In the case of $\frac{n}{2} < k < n, n-k < k$. Given that $i+1 \leq n-k+1$, then $i+1 \leq n-k+1 \leq k$. The reliability is $R_{opt}^t(i) =$ and $R_{opt}^t(i-1) = 1$. Thus, the component integrated importance measure be expressed as $I_i^{im}(t) = P(t)\lambda_i(t)\frac{R_{opt}^t(n-i)-R_{opt}^t(n)}{1-P(t)}$ and $I_{i+1}^{im}(t) = P(t)\lambda_{i+1}(t)\frac{R_{opt}^t(n-i-1)-R_{opt}^t(n)}{1-P(t)}$. Given that $R_{opt}^t(n-i) < R_{opt}^t(n-i-1)$, then we can obtain $I_i^{im}(t) \leq I_{i+1}^{im}(t)$ when $\lambda_i(t) \leq \lambda_{i+1}(t)$.

In the case of $2 < k \leq \frac{n}{2}, k \leq n-k$. Given that $i > n-k$, then $k \leq n-k < i$. Furthermore, $R_{opt}^t(n-i) =$ and $R_{opt}^t(n-i-1) = 1$. Therefore, the component integrated importance measure can be expressed as $I_i^{im}(t) = P(t)\lambda_i(t)\frac{R_{opt}^t(i-1)-R_{opt}^t(n)}{1-P(t)}$ and $I_{i+1}^{im}(t) = P(t)\lambda_{i+1}(t)\frac{R_{opt}^t(i)-R_{opt}^t(n)}{1-P(t)}$. Given that $R_{opt}^t(i) < R_{opt}^t(i-1)$, then we can obtain $I_i^{im}(t) \geq I_{i+1}^{im}(t)$ when $\lambda_i(t) \geq \lambda_{i+1}(t)$.

In the case of $2 < k \leq \frac{n}{2}, k \leq n-k$. Given that $i < k$, we have $i < k \leq n-k$. Subsequently, the reliability is $R_{opt}^t(i) =$ and $R_{opt}^t(i-1) = 1$. The component integrated importance measure is $I_i^{im}(t) = P(t)\lambda_i(t)\frac{R_{opt}^t(n-i)-R_{opt}^t(n)}{1-P(t)}$ and $I_{i+1}^{im}(t) = P(t)\lambda_{i+1}(t)\frac{R_{opt}^t(n-i-1)-R_{opt}^t(n)}{1-P(t)}$. Given that $R_{opt}^t(n-i) < R_{opt}^t(n-i-1)$, then $I_i^{im}(t) \leq I_{i+1}^{im}(t)$ can be obtained when $\lambda_i \leq \lambda_{i+1}$. □

According to Propositions 2.1 and 2.2, if the failure rate function of component i is less than that of adjacent component i+1, the impact of component i on system performance is less than that of component i+1. This situation also means that when considering the impact of components on system reliability, the most important components can be identified. In addition, a greater increase in system reliability would require significant improvements in components with large importance values.

2.3 Joint Importance Measures for Reliability Design

Marginal reliability importance measures quantify mainly the change in system reliability and the relationship with the change in reliability of a component of the system

[14, 28]. Hong et al. [16] studied the joint reliability importance (JRI) of components in a k-out-of-n: G system. Gao et al. [15] extended the concept of JRI and JFI (joint failure importance) of two components to a multi-component system and further discussed some relationships between their JRI and JRI, JFI and JFI, and JFI and JRI, respectively. Rani et al. [35] studied conditional marginal and conditional joint reliability importance in series-parallel systems. Eryilmaz [13] studied joint reliability importance in a linear m-consecutive-k-out-of-n: F system.

2.3.1 The Calculation of Joint Reliability Importance in k-out-of-n: F Systems

The JRI measures how much influence the interaction between components has on the reliability of a system. On the promise that components are statistically independent and identical distributed, the joint reliability importance of component pair $(i,\ j)$ in a k-out-of-n: F system is defined by:

$$I_{i,j}^{j} = \frac{\partial^2 R(p,k,n)}{\partial p_i \partial p_j}, \tag{2.100}$$

where P_i and P_j are reliabilities of components i and j, respectively, and $\boldsymbol{P} = (P_1, P_2, \ldots, P_n)$ is the reliability vector of all components.

If the components of the system are independently and identically distributed, the reliability of a k-out-of-n: F system is:

$$R(p,n,k) = \sum_{r=k}^{n} C_n^r p^r q^{n-r}. \tag{2.101}$$

When the components in the system are independently and identically distributed,

$$\begin{aligned} I_{i,j}^{j} &= \frac{\partial^2 R(p,k,n)}{\partial p^2} \\ &= R(p,k,n-2) + R(p,k-2,n-2) - 2R(p,k-1,n-2). \end{aligned} \tag{2.102}$$

It can be seen from Eq. (2.102) that when the component i and component j are both in working states, the reliability of system $R(p,k,n-2)$ is

$$R(p,k,n-2) = \sum_{r=k}^{n-2} C_{n-2}^r p^r q^{n-r-2}. \tag{2.103}$$

In the same way, when the component i and component j are both in failed states, the reliability of the system is given by

$$R(p, k-2, n-2) = \sum_{r=k-2}^{n-2} C_{n-2}^{r} p^{r} q^{n-r-2}. \tag{2.104}$$

When component i is in a failed state while component j is in a working state, the reliability of system $R(p, k, n-2)$ is

$$R(p, k-1, n-2) = \sum_{r=k-1}^{n-2} C_{n-2}^{r} p^{r} q^{n-r-2}. \tag{2.105}$$

Substituting Eqs. (2.103), (2.104), and (2.105) into Eq. (2.102), we can obtain

$$I_{i,j}^{j} = C_{n-2}^{k-2} p^{k-2} q^{n-k-2} - C_{n-2}^{k-1} p^{k-1} q^{n-k-1}. \tag{2.106}$$

By further simplification, we obtain

$$I_{i,j}^{j} = p^{k-2} q^{n-k-1} (C_{n-2}^{k-2} - p C_{n-1}^{k-1}). \tag{2.107}$$

2.3.2 Analysis for the Relevant Properties of the Joint Reliability Importance in k-out-of-n: F Systems

The parameters (i.e., k and n) of a k-out-of-n: F system will affect the calculation results of the joint reliability importance. Here we will analyse the main parameters p, k and n that affect the result.

Remark 2.2 There always exists a certain value of p, which makes the value of JRI(i, j) in an k-out-of-n: F system be 0, and $p=(k\text{-}1)/(n\text{-}1)$.

Next, we will conduct relevant analysis and derivation based on Eq. (2.107).

- If $p = (k-1)/(n-1)$, $I_{i,j}^{j} = 0$. In this case, component i has the same effect on the reliability of the system $R(p, n, k)$ no matter whether component j is in a working state or the failed state.
- If $0 < p < (k-1)/(n-1)$, $I_{i,j}^{j}$ >0. In this case, the effect of component i on system reliability when component j is in a working state is larger than the effect of component i on system reliability when component j is in the failed state.
- If $(k-1)/(n-1) < p < 1$, $I_{i,j}^{j}$ <0. In this case, the effect of component i on system reliability when component j is in the failed state is larger than the effect of component i on system reliability when component j is in a working state.

Remark 2.3 We make the following comments.

- The value of $I_{i,j}^{j}$ in a k-out-of-n: F system first increases rapidly to the maximum, then decreases to the minimum rapidly. Next, it increases quickly once again and approaches 0 infinitely. After that, the value of $I_{i,j}^{j}$ remains unchanged.

- The value of $I_{i,j}^{j}$ in a k-out-of-n: F system decreases with the increase of the value of k monotonically.
- The value of $I_{i,j}^{j}$ in a k-out-of-n: F system increases to the maximum rapidly and then decreases rapidly to the minimum with the increase of the value of n. Next, it increases rapidly once again and approaches 0 infinitely. After that, it remained unchanged generally.

2.3.3 The Calculation and Analysis of Joint Reliability Importance in Consecutive k-out-of-n: F System

Joint reliability importance measures need to be discussed according to different system structures. Therefore, we need to perform the derivation and analysis of the joint reliability importance for linear and circular consecutive k-out-of-n: F systems, respectively. Below we only perform the derivation of the joint importance measure for a linear consecutive k-out-of-n: F system.

In the following, we will take the joint reliability importance of a component pair (i, j) in a linear consecutive k-out-of-n: F system as an example to derive. On the premise that components in the system are statistically independent and identical distributed, we further derive the basic expression of the joint reliability importance in a consecutive k-out-of-n: F system:

$$I_{i,j}^{j} = \frac{P\{\mathbf{W}\} - P\{\mathbf{W}, Y_i = 0\} - P\{\mathbf{W}, Y_j = 0\} + P\{\mathbf{W}, Y_i = 0, Y_j = 0\}}{1 - P\{Y_j = 0\} - P\{Y_i = 0\} + P\{Y_i = 0, Y_j = 0\}} - \frac{[P\{\mathbf{W}, Y_j = 0\} - P\{\mathbf{W}, Y_i = 0, Y_j = 0\}]}{P\{Y_j = 0\} - P\{Y_i = 0, Y_j = 0\}} - \frac{[P\{\mathbf{W}, Y_i = 0\} - P\{\mathbf{W}, Y_i = 0, Y_j = 0\}]}{P\{Y_i = 0\} - P\{Y_i = 0, Y_j = 0\}} + \frac{P\{\mathbf{W}, Y_i = 0, Y_j = 0\}}{P\{Y_i = 0, Y_j = 0\}}. \tag{2.108}$$

Thus, for the computation of the joint reliability importance of component pair (i, j), we need to evaluate the following two probabilities: $P\{\mathbf{W}, Y_i = 0\}$ and $P\{\mathbf{W}, Y_i = 0, Y_j = 0\}$.

For a consecutive k-out-of-n: F system consists of n independent components with common working probability $P\{Y_i = 0\} = p$, we obtain

$$P\{W\} = P\{N_{n,k} < 1\} = 1 - \sum_{s=1}^{[\frac{n}{k}]} \sum_{i=sk}^{n} C(i, s, k, n) p^{n-i} q^{i}, \tag{2.109}$$

where

$$C(i, s, k, n) = C_{s+n-i}^{s} \sum_{j=0}^{\min(n-i+1,[\frac{i-ks}{k}])} (-1)^j C_{n-i+1}^{j} C_{n-ks-jk}^{n-i}. \tag{2.110}$$

and

$$\begin{aligned} P\{W, Y_i = 0\} &= P\left\{N_{n,k} < m, Y_i = 0\right\} \\ &= \sum_{s_1=0}^{u} \sum_{s_2=0}^{v} \sum_{m_1=0}^{i-1} \sum_{m_2=0}^{n-i} C(i - m_1 - 1, s_1, k, i - 1). \end{aligned} \tag{2.111}$$

Thus, we can obtain

$$\begin{aligned} P\{W, Y_i = 0, Y_j = 0\} = \sum_{s_1} \sum_{s_2} \sum_{s_3} \sum_{m_1=0}^{i-1} \sum_{m_2=0}^{j-i-1} \sum_{m_3=0}^{n-j} P\{N_{i-1,k}^{(1,i-1)} = s_1, \\ N_{j-i-1,k}^{(i+1,j-1)} = s_2, N_{n-j}^{(j+1,n)} = s_3, S_{i-1}^{(1,i-1)} = m_1 N_{j-i-1,k}^{(i+1,j-1)} = m_2, \\ S_{n-j}^{(j+1,n)} = m_3, Y_i = 0, Y_j = 0\}. \end{aligned} \tag{2.112}$$

Conditioning on the number of working components, we can rewrite Eq. (2.112) as:

$$\begin{aligned} P\{W, Y_i = 0, Y_j = 0\} = P\{N_{i-1,k}^{(1,i-1)} + N_{j-i-1,k}^{(i-1,i-1)} + N_{n-j,k}^{(j+1,n)}, Y_i = 0, Y_j = 0\} \\ = \sum_{s_1} \sum_{s_2} \sum_{s_3} \Pr\{N_{i-1,k}^{(1,i-1)} = s_1, N_{j-i-1,k}^{(i-1,i-1)} = s_2, N_{n-j,k}^{(j+1,n)} = s_3, Y_i = 0, Y_j = 0\}. \end{aligned} \tag{2.113}$$

The components are independent, so

$$\begin{aligned} &P\{N_{i-1,k}^{(1,i-1)} = s_1, N_{j-i-1,k}^{(i+1,j-1)} = s_2, N_{n-j,k}^{(j+1,n)} = s_3, S_{i-1}^{(1,i-1)} = m_1, S_{j-i-1}^{(i+1,j-1)} = m_2, \\ &S_{n-j}^{(j+1,n)} = m_3, Y_i = 0, Y_j = 0\} \\ &= P\{N_{i-1,k}^{(1,i-1)} = s_1, S_{i-1}^{(1,i-1)} = m_1\} P\{N_{j-i-1,k}^{(i+1,j-1)} = s_2, S_{j-i-1}^{(i+1,j-1)} = m_2\} \\ &P\{N_{n-j,k}^{(j+1,n)} = s_3, S_{n-j}^{(j+1,n)} = m_3\} P\{Y_i = 0\} P\{Y_j = 0\}. \end{aligned} \tag{2.114}$$

Based on Eq. (2.111), we can obtain

$$P\{N_{i-1,k}^{(1,i-1)} = s_1, S_{i-1}^{(i,i-1)} = m_1\} = C(i - m_1 - 1, s_1, k, i - 1) p^{m_1} q^{i-m_1-1}. \tag{2.115}$$

$$
\begin{aligned}
P\{N_{i-1,k}^{(1,i-1)} &= s_1, N_{j-i-1,k}^{(i+1,j-1)} = s_2, N_{n-j,k}^{(j+1,n)} = s_3, S_{i-1}^{(1,i-1)} = m_1, S_{j-i-1}^{(i+1,j-1)} = m_2, \\
&S_{n-j}^{(j+1,n)} = m_3, Y_i = 0, Y_j = 0\} \\
&= C(i - m_1 - 1, s_1, k, i - 1)C(j - i - m_2 - 1, s_2, k, j - i - 1) \\
&\quad C(n - j - m_3, s_3, k, n - j)p^{m_1+m_2+m_3+2}q^{n-m_1-m_2-m_3-2}. \qquad (2.116)
\end{aligned}
$$

Substituting Eq. (2.116) into Eq. (2.113), we can obtain

$$
\begin{aligned}
&P\{\mathbf{W}, Y_i = 0, Y_j = 0\} \\
&= \sum_{s_1=0}^{u}\sum_{s_2=0}^{y}\sum_{s_3=0}^{z}\sum_{m_1=0}^{i-1}\sum_{m_2=0}^{j-i-1}\sum_{m_3=0}^{n-j}[C(i - m_1 - 1, s_1, k, i - 1) \\
&\quad \times C(j - i - m_2 - 1, s_2, k, j - i - 1) \\
&\quad \times C(n - j - m_3, s_3, k, n - j)p^{m_1+m_2+m_3+2}q^{n-m_1-m_2-m_3-2}\Big], \qquad (2.117)
\end{aligned}
$$

where $u = \min\{ m - 1, [\frac{i-1}{k}]\}$, $y = \min\{ m - s_1 - 1, [\frac{j-i-1}{k}]\}$, $z = \min\{ m - s_1 - s_2 - 1, [\frac{n-j}{k}]\}$.

Substituting Eqs. (2.111) and (2.117) in Eq. (2.108), we can obtain the expression of the joint reliability importance in a linear consecutive k-out-of-n: F system as

$$
\begin{aligned}
I_{i,j}^{j} &= \frac{1}{1 - 2p + p^2}\Bigg\{1 - \sum_{s=1}^{[\frac{n}{k}]}\sum_{i=sk}^{n} C(i, s, k, n)p^{n-i}q^{i} \\
&\quad - \sum_{s_1=0}^{u}\sum_{s_2=0}^{v}\sum_{m_1=0}^{i-1}\sum_{m_2=0}^{n-i} C(i - m_1 - 1, s_1, k, i - 1) \\
&\quad - \sum_{s_1=0}^{u}\sum_{s_2=0}^{v}\sum_{m_1=0}^{j-1}\sum_{m_2=0}^{n-j} C(j - m_1 - 1, s_1, k, j - 1) \\
&\quad + \sum_{s_1=0}^{u}\sum_{s_2=0}^{y}\sum_{s_3=0}^{z}\sum_{m_1=0}^{i-1}\sum_{m_2=0}^{j-i-1}\sum_{m_3=0}^{n-j}[C(i - m_1 - 1, s_1, k, i - 1) \\
&\quad C(j - i - m_2 - 1, s_2, k, j - i - 1)C(n - j - m_3, s_3, k, n - j) \\
&\quad p^{m_1+m_2+m_3+2}q^{n-m_1-m_2-m_3-2}\Big]\Bigg\} \\
&\quad - \frac{1}{p - p^2}\Bigg\{\sum_{s_1=0}^{u}\sum_{s_2=0}^{v}\sum_{m_1=0}^{i-1}\sum_{m_2=0}^{n-i} C(i - m_1 - 1, s_1, k, i - 1) \\
&\quad + \sum_{s_1=0}^{u}\sum_{s_2=0}^{v}\sum_{m_1=0}^{j-1}\sum_{m_2=0}^{n-j} C(j - m_1 - 1, s_1, k, j - 1)
\end{aligned}
$$

$$-2\sum_{s_1=0}^{u}\sum_{s_2=0}^{y}\sum_{s_3=0}^{z}\sum_{m_1=0}^{i-1}\sum_{m_2=0}^{j-i-1}\sum_{m_3=0}^{n-j} C(i-m_1-1, s_1, k, i-1)$$
$$C(j-i-m_2-1, s_2, k, j-i-1)C(n-j-m_3, s_3, k, n-j)$$
$$\left. p^{m_1+m_2+m_3+2} q^{n-m_1-m_2-m_3-2} \right\}$$
$$+\frac{1}{p^2}\left\{\sum_{s_1=0}^{u}\sum_{s_2=0}^{y}\sum_{s_3=0}^{z}\sum_{m_1=0}^{i-1}\sum_{m_2=0}^{j-i-1}\sum_{m_3=0}^{n-j} C(i-m_1-1, s_1, k, i-1)\right.$$
$$C(j-i-m_2-1, s_2, k, j-i-1)C(n-j-m_3, s_3, k, n-j)$$
$$\left. p^{m_1+m_2+m_3+2} q^{n-m_1-m_2-m_3-2} \right\}. \tag{2.118}$$

2.4 Joint Importance Measures for System Reconfiguration

It is known that the reliability of a system can be improved by adding more redundant components. For example, in a k-out-of-n: F system, the system will fail if k components fail [26].

There is often a need to optimise a continuous k-out-of-n system in terms of seeking the optimal allocation of components to maximise the reliability of the system. Levitin [21] discussed the optimisation of a linear polymorphic sliding window system. Yam et al. [41] studied the reliability of a repairable circular continuous k-out-of-n: F system. Jalali et al. [18] provided a proof of the optimal arrangement for a linear continuous k-out-of-n system with the optimal arrangement was proved. Mo et al. [31] analysed the performance of a multi-state continuous k-out-of-n: F system. Zhu et al. [45] proposed a Birnbaum importance-based local search method and genetic algorithm to solve the component assignment problem.

2.4.1 *Joint Integrated Importance Measure (JIIM)*

The joint failure importance of components i and l can be denoted as [47]

$$I_{i,l}^{j} = R\left(1_i, 1_l, \boldsymbol{p}(\boldsymbol{t})\right) + R\left(0_i, 0_l, \boldsymbol{p}(\boldsymbol{t})\right) - R\left(1_i, 0_l, \boldsymbol{p}(\boldsymbol{t})\right) - R\left(0_i, 1_l, \boldsymbol{p}(\boldsymbol{t})\right). \tag{2.119}$$

Then in a unit of time, the change of the system reliability made by repairing component i is

$$\begin{aligned}
&\frac{d\left[R\left(1_i, p\right)-R\left(0_i, p\right)\right]}{dt} \\
&= \sum_{l=1, l\neq i}^{n} \frac{dR_l(t)}{dt} \frac{\partial R\left(1_i, p\right)}{\partial R_l(t)} - \sum_{l=1, l\neq i}^{n} \frac{dR_l(t)}{dt} \frac{\partial R\left(0_i, p\right)}{\partial R_l(t)} \\
&= \sum_{l=1, l\neq i}^{n} \frac{dR_l(t)}{dt}\left[R\left(1_i, 1_l, p\right)-R\left(1_i, 0_l, p\right)\right] \\
&\quad - \sum_{l=1, l\neq i}^{n} \frac{dR_l(t)}{dt}\left[R\left(0_i, 1_l, p\right)-R\left(0_i, 0_l, p\right)\right] \\
&= \sum_{l=1, l\neq i}^{n} \frac{dR_l(t)}{dt}\left[R\left(1_i, 1_l, p\right)+R\left(0_i, 0_l, p\right)-R\left(1_i, 0_l, p\right)-R\left(0_i, 1_l, p\right)\right] \\
&= -\sum_{l=1, l\neq i}^{n} \lambda_l(t) R_l(t)\left[R\left(1_i, 1_l, p\right)+R\left(0_i, 0_l, p\right)-R\left(1_i, 0_l, p\right)-R\left(0_i, 1_l, p\right)\right] \\
&= -\sum_{l=1, l\neq i}^{n} \lambda_l(t) R_l(t) I_{i,l}^{j}.
\end{aligned} \tag{2.120}$$

Equation (2.120) suggests that when component i is repaired, the change of system reliability in a unit of time is the sum of $\lambda_l(t)R_l(t)I_{i,l}^{j}(t)$ of all other system component l, in which $\lambda_l(t)R_l(t)I_{i,l}^{j}(t)$ is the contribution of component l to the change of the system reliability in a unit of time after component i is repaired. $R_i(t)\lambda_i(t)[R\left(1_i, \boldsymbol{p}(t)\right) - R(0_i, \boldsymbol{p}(t))]$ of component i represents the change of system reliability in a unit of time due to component i. Thus, the joint integrated importance between components i and l is $I_{i,l}^{jim}(t) = \lambda_i(t)\lambda_l(t)R_i(t)R_l(t)I_{i,l}^{j}(t)$, which represents the joint contribution between components i and l to the change of system reliability in a unit of time. For component i, when the effect of the sum of all other system component l is considered, we have $I_{i,l}^{jim}(t) = \sum_{l=1,l\neq i}^{n} \lambda_i(t)\lambda_l(t)R_i(t)R_l(t)I_{i,l}^{j}(t)$.

2.4.2 *Joint Differential Importance Measure (JDIM)*

The Differential Importance Measure (DIM) is defined by Borgonovo and Apostolakis [4]

$$I_i^{\mathrm{D}} = \frac{\frac{\partial R}{\partial X_i} dX_i}{\sum_j \frac{\partial R}{\partial X_j} dX_j} = \frac{I_i^B dX_i}{\sum_j I_i^B dX_j}. \tag{2.121}$$

A change in system reliability due to arbitrary changes in the values of all component reliabilities can be expressed as

$$\Delta R = \sum_{i=1}^{n} \frac{\partial R}{\partial R_i} \Delta R_i + \sum_{i=1}^{n} \sum_{l>i=1}^{n} \frac{\partial^2 R}{\partial R_i \partial R_l} \Delta R_i \Delta R_l$$
$$+ \sum_{i=1}^{n} \sum_{l>i=1}^{n} \sum_{h>l=1}^{n} \frac{\partial^3 R}{\partial R_i \partial R_l \partial R_h} \Delta R_i \Delta R_l \Delta R_h + \ldots \tag{2.122}$$

Considering the second order Taylor expansion, the system reliability change is

$$\Delta R^{\mathrm{II}} = \sum_{i=1}^{n} \frac{\partial R}{\partial R_i} \Delta R_i + \sum_{i=1}^{n} \sum_{l>i=1}^{n} \frac{\partial^2 R}{\partial R_i \partial R_l} \Delta R_i \Delta R_l. \tag{2.123}$$

With reference to components i and l, the variation of system reliability R, ΔR_{il}, due to the variations of the component reliabilities ΔR_i and ΔR_l is

$$\Delta R_{il}^{\mathrm{II}} = \frac{\partial R}{\partial R_i} \Delta R_i + \frac{\partial R}{\partial R_l} \Delta R_l + \frac{\partial^2 R}{\partial R_i \partial R_l} \Delta R_i \Delta R_l. \tag{2.124}$$

The joint DIM (JDIM) of components i and l is (a second-order DIM):

$$I_{il}^{\mathrm{D}} = \frac{\Delta R_{il}^{\mathrm{II}}}{\Delta R^{\mathrm{II}}} = \frac{\frac{\partial R}{\partial R_i} \Delta R_i + \frac{\partial R}{\partial R_l} \Delta R_l + \frac{\partial^2 R}{\partial R_i \partial R_l} \Delta R_i \Delta R_l}{\sum_{i=1}^{n} \frac{\partial R}{\partial R_i} \Delta R_i + \sum_{i=1}^{n} \sum_{l>i=1}^{n} \frac{\partial^2 R}{\partial R_i \partial R_l} \Delta R_i \Delta R_l}. \tag{2.125}$$

For component i,

$$\Delta R_{i}^{\mathrm{II}} = \frac{\partial R}{\partial R_i} \Delta R_i + \sum_{l=1, l\neq i}^{n} \frac{\partial^2 R}{\partial R_i \partial R_l} \Delta R_i \Delta R_l, \tag{2.126}$$

and

$$I_{i}^{\mathrm{D}} = \frac{\Delta R_{i}^{\mathrm{II}}}{\Delta R^{\mathrm{II}}} = \frac{\frac{\partial R}{\partial R_i} \Delta R_i + \sum_{l=1, l\neq i}^{n} \frac{\partial^2 R}{\partial R_i \partial R_l} \Delta R_i \Delta R_l}{\sum_{i=1}^{n} \frac{\partial R}{\partial R_i} \Delta R_i + \sum_{i=1}^{n} \sum_{l>i=1}^{n} \frac{\partial^2 R}{\partial R_i \partial R_l} \Delta R_i \Delta R_l}. \tag{2.127}$$

By this definition, I_{il}^{D} captures second order interaction effects in determining component i importance.

2.4.3 Binary Systems

Denote $R_{i\to j} = R\left([i] \to [j]\right)$ and $R_{i\to j}^{'} = R^{'}\left([i] \to [j]\right)$. By using the definition of the consecutive-k-out-of-n: F system, we have

$$
\begin{aligned}
&R\left(1_i, \boldsymbol{p}(\boldsymbol{t})\right) \\
&= R\left(p_1, \ldots, p_{i-1}, 1, p_{i+1}, \ldots, p_n\right) \\
&= R\left(p_1, \ldots, p_{i-1}\right) R\left(p_{i+1}, \ldots, p_n\right) \\
&= R_{1\to i-1} R_{i+1\to n} R\left(0_i, \boldsymbol{p}(\boldsymbol{t})\right) \\
&= R\left(p_1, \ldots, p_{i-1}, 0, p_{i+1}, \ldots, p_n\right) \\
&= R_{1\to i-1} R\left(p_{i-k+1}, \ldots, p_{i-1}, 0, p_{i+1}, \ldots, p_{i+k-1}\right) R_{i+1\to n} \\
&= R_{1\to i-1} R'_{i-k+1\to i+k-1} R_{i+1\to n},
\end{aligned}
\tag{2.128}
$$

and

$$
R(n) = p_i R\left(1_i, \boldsymbol{p}(\boldsymbol{t})\right) + q_i R\left(0_i, \boldsymbol{p}(\boldsymbol{t})\right). \tag{2.129}
$$

By using the definition of the Birnbaum importance, we can obtain

$$
\begin{aligned}
R(I_i^{\mathrm{B}}) &= R\left(1_i, \boldsymbol{p}(\boldsymbol{t})\right) - R\left(0_i, \boldsymbol{p}(\boldsymbol{t})\right) \\
&= \frac{R_{1\to i-1} R_{i+1\to n} - R(n)}{q_i}.
\end{aligned}
\tag{2.130}
$$

Imposing $i + 1 < j$, we can derive the following definition:

$$
R\left(1_i, 1_l, \boldsymbol{p(t)}\right) = R_{1\to i-1} R_{i+1\to l-1} R_{l+1\to n}, \tag{2.131}
$$

and

$$
\begin{aligned}
&R\left(1_i, 0_l, \boldsymbol{p(t)}\right) = R\left(p_1, \ldots, p_{i-1}, 1, p_{i+1}, \ldots, p_{l-1}, 0, p_{l+1}, \ldots, p_n\right) \\
&= R\left(p_1, \ldots, p_{i-1}\right) R\left(p_{i+1}, \ldots, p_{l-1}, 0, p_{l+1}, \ldots, p_n\right) \\
&= R_{1\to i-1} R_{i+1\to l-1} R\left(p_{l-k+1}, \ldots, p_{l-1}, 0, p_{l+1}, \ldots, p_{l+k-1}\right) \\
&\ R\left(l+1 \to n\right) \\
&= R_{1\to i-1} R_{i+1\to l-1} R'_{i-k+1\to i+k-1} R_{i+1\to n}.
\end{aligned}
\tag{2.132}
$$

Let $h_{q,p}(t) = (q_i(t)p_l(t) - p_i(t)p_l(t))$. Since Eq. (2.132) helps us to derive the specific expansion of $I_{i,l}^{j0}$, we can obtain,

$$
\begin{aligned}
&R\left(1_i, 1_l, \boldsymbol{p(t)}\right) - R\left(0_i, 1_l, \boldsymbol{p(t)}\right) \\
&= \frac{1}{q_i(t)p_l(t)} \left[h_{q,p}(t) R_{1\to i-1} R_{i+1\to l-1} R_{i+1\to n} + R\left(1_l, \boldsymbol{p(t)}\right)\right] \\
&= \frac{\left(q_i(t)p_l(t) - p_i(t)p_l(t)\right) R_{1\to i-1} R_{i+1\to l-1} R_{i+1\to n} + R_{1\to l-1} R_{i+1\to n}}{q_i(t)p_l(t)} \\
&= \frac{\left\{h_{q,p}(t) R_{1\to i-1} R_{i+1\to l-1} + R_{1\to l-1}\right\} R_{i+1\to n}}{q_i(t)p_l(t)},
\end{aligned}
\tag{2.133}
$$

and

$$
\begin{aligned}
& R\left(1_i, 0_l, \boldsymbol{p(t)}\right) - R\left(0_i, 0_l, \boldsymbol{p(t)}\right) \\
&= \frac{\left(q_i(t)q_l(t) - p_i(t)q_l(t)\right) R_{1\to i-1} R_{i+1\to l-1} R^{'}_{l-k+1\to l+k-1} R_{i+1\to n} + R\left(0_l, \boldsymbol{p(t)}\right)}{q_i(t)q_l(t)} \\
&= \frac{\left(q_i(t)q_l(t) - p_i(t)q_l(t)\right) R_{1\to i-1} R_{i+1\to l-1} R^{'}_{l-k+1\to l+k-1} R_{i+1\to n}}{q_i(t)q_l(t)} \\
&\quad + \frac{R_{1\to l-1} R^{'}_{l-k+1\to l+k-1} R_{i+1\to n}}{q_i(t)q_l(t)} \\
&= \frac{\left\{\left(q_i(t)q_l(t) - p_i(t)q_l(t)\right) R_{1\to i-1} R_{i+1\to l-1} + R_{1\to l-1}\right\} R^{'}_{l-k+1\to l+k-1} R_{i+1\to n}}{q_i(t)q_l(t)}.
\end{aligned}
\tag{2.134}
$$

Then we have

$$
\begin{aligned}
& R(I^j_{i,l}) \\
&= R\left(1_i, 1_l, \boldsymbol{p(t)}\right) - R\left(0_i, 1_l, \boldsymbol{p(t)}\right) - \left(R\left(1_i, 0_l, \boldsymbol{p(t)}\right) - R\left(0_i, 0_l, \boldsymbol{p(t)}\right)\right) \\
&= \frac{\left\{h_{q,p}(t) R_{1\to i-1} R_{i+1\to l-1} + R_{1\to l-1}\right\} R_{i+1\to n}}{q_i(t)p_l(t)} \\
&\quad - \frac{\left\{\left(q_i(t)q_l(t) - p_i(t)q_l(t)\right) R_{1\to i-1} R_{i+1\to l-1} + R_{i\to l-1}\right\} R^{'}_{l-k+1\to l+k-1} R_{l+1\to n}}{q_i(t)q_l(t)} \\
&= \frac{q_i(t)q_l(t)\left\{h_{q,p}(t) R_{1\to i-1} R_{i+1\to l-1} + R_{i\to l-1}\right\} R_{l+1\to n}}{q_i(t)p_l(t)q_i(t)q_l(t)} \\
&\quad - \frac{R^{'}_{l-k+1\to l+k-1} q_i(t)p_l(t)\left\{\left(q_i(t)q_l(t) - p_i(t)q_l(t)\right) R_{1\to i-1} R_{i+1\to l-1} + R_{i\to l-1}\right\} R_{l+1\to n}}{q_i(t)q_l(t)q_i(t)p_l(t)} \\
&= \frac{\left\{\begin{array}{c} \left\{q_i(t)q_l(t)h_{q,p}(t) - q_i(t)p_l(t)\left(q_i(t)q_l(t) - p_i(t)q_l(t)\right) R^{'}_{l-k+1\to l+k-1}\right\} \\ R_{1\to i-1} R_{i+1\to l-1} + \left\{q_i(t)q_l(t) - q_i(t)p_l(t) R^{'}_{l-k+1\to l+k-1}\right\} R_{i\to l-1} + Rl + 1 \to n \end{array}\right\}}{q_i^2(t)p_l(t)q_l(t)}.
\end{aligned}
\tag{2.135}
$$

With component i, when considering the effect of the sum of all other system component l, the joint integrated importance for the consecutive-k-out-of-n: F system is as follows:

$$
I^j_{i,l} = \sum_{l=1, l\neq i}^{n} \lambda_i(t)\lambda_l(t) R_i(t) R_l(t) R\left(I^j_{i,l}\right). \tag{2.136}
$$

2.4.4 Multistate Systems

In order to enhance its applicability, we extend the two joint importance measures of reconfigurable systems to multistate systems. We stipulate $R_i\,(v) = \Pr\,[x_i \geq v]$, $R\,(v_i, \boldsymbol{p}; m) = \Pr\left[\phi(v_i, \boldsymbol{p}) \geq m\right]$. Then,

$$
\begin{aligned}
&\frac{d\left[R\,(v_i, \boldsymbol{p}) - R\,((v-1)_i, \boldsymbol{p})\right]}{dt} \\
&= \sum_{m=1}^{M}\left\{\sum_{l=1,l\neq i}^{n}\frac{dR_l(k)}{dt}\frac{\partial R\,(v_i, \boldsymbol{p}; m)}{\partial R_l(k)} - \sum_{l=1,l\neq i}^{n}\frac{dR_l(k)}{dt}\frac{\partial R\,((v-1)_i, \boldsymbol{p}; m)}{\partial R_l(k)}\right\} \\
&= \sum_{m=1}^{M}\left\{\sum_{l=1,l\neq i}^{n}\frac{dR_l(k)}{dt}\left[R\,(v_i, k_l, \boldsymbol{p}; m) - R\,(v_i, (k-1)_l, \boldsymbol{p}; m)\right]\right. \\
&\quad \left. - \sum_{l=1,l\neq i}^{n}\frac{dR_l(k)}{dt}\left[R\,((v-1)_i, k_l, \boldsymbol{p}; m) - R\,((v-1)_i, (k-1)_l, \boldsymbol{p}; m)\right]\right\} \\
&= \sum_{m=1}^{M}\left\{\sum_{l=1,l\neq i}^{n}\frac{dR_l(k)}{dt}\left[R\,(v_i, k_l, \boldsymbol{p}; m) + R\,((v-1)_i, (k-1)_l, \boldsymbol{p}; m)\right.\right. \\
&\quad \left.\left. - R\,(v_i, (k-1)_l, \boldsymbol{p}; m) - R\,((v-1)_i, k_l, \boldsymbol{p}; m)\right]\right\} \\
&= -\sum_{l=1,l\neq i}^{n}\sum_{m=1}^{M}\lambda_l(k)R_l(k)\left[R\,(v_i, k_l, \boldsymbol{p}; m) + R\,((v-1)_i, (k-1)_l, \boldsymbol{p}; m)\right. \\
&\quad \left. - R\,(v_i, (k-1)_l, \boldsymbol{p}; m) - R\,((v-1)_i, k_l, \boldsymbol{p}; m)\right].
\end{aligned}
\tag{2.137}
$$

We can obtain,

$$
\begin{aligned}
R^m(I^j_{i_v,l_k}) = \sum_{m=1}^{M}&\left[R\,(v_i, k_l, \boldsymbol{p}; m) + R\,((v-1)_i, (k-1)_l, \boldsymbol{p}; m)\right. \\
&\left. - R\,(v_i, (k-1)_l, \boldsymbol{p}; m) - R\,((v-1)_i, k_l, \boldsymbol{p}; m)\right].
\end{aligned}
\tag{2.138}
$$

The Birnbaum importance measure in multistate systems can be expressed as

$$
R^m(I^B_{i_v}) = \sum_{m=1}^{M}[R\,(v_i, \boldsymbol{p}; m) - R\,((v-1)_i, \boldsymbol{p}; m)]. \tag{2.139}
$$

For linear consecutive-k-out-of-n systems, we assume that v_0 is the threshold state of component i, and k_0 is the threshold state of component l. Once the component state is degraded below the threshold state, it will fail. Similarly, we can obtain

$$
\begin{aligned}
&R\left((\geq v_0)_i, (\geq k_0)_l, \boldsymbol{p}(t); m\right) \\
&= R_{1\to i-1} R_{i+1\to l-1} R_{l+1\to n}, R\left((\geq v_0)_i, (< k_0)_l, \boldsymbol{p}(t); m\right) \\
&= R_{1\to i-1} R_{i+1\to l-1} R\left((\geq v_0)_{l-k+1}, \ldots, (\geq v_0)_{l-1}, (< k_0)_l, (\geq v_0)_{l+1}, \ldots, (\geq v_0)_{l+k-1}\right) R_{l+1\to n} \\
&= R_{1\to i-1} R_{i+1\to l-1} R'_{i-k+1\to i+k-1} R_{i+1\to n}, R\left((\geq v_0)_l, \boldsymbol{p}(t)\right) \\
&= R_i(\geq v_0) R_l(\geq k_0) R\left((\geq v_0)_i, (\geq k_0)_l, \boldsymbol{p}(t)\right) \\
&\quad + R_i(< v_0) R_l(\geq k_0) R\left((< v_0)_i, (\geq k_0)_l, \boldsymbol{p}(t)\right), R\left((< v_0)_l, \boldsymbol{p}(t)\right) \\
&= R_i(\geq v_0) R_l(< k_0) R\left((\geq v_0)_i, (< k_0)_l, \boldsymbol{p}(t)\right) \\
&\quad + R_i(< v_0) R_l(< k_0) R\left((< v_0)_i, (< k_0)_l, \boldsymbol{p}(t)\right).
\end{aligned}
\tag{2.140}
$$

According to Eq. (2.140), for different states of components, we can calculate the corresponding $R(v_i, k_l, \boldsymbol{p}; m) + R((v-1)_i, (k-1)_l, \boldsymbol{p}; m) - R(v_i, (k-1)_l, \boldsymbol{p}; m) - R((v-1)_i, k_l, \boldsymbol{p}; m)$.

Assume that $R^m_{opt}(I^B_{i_v})$ and $R^m_{opt}(I^j_{i_v,l_k})$ are the reliability of systems or subsystems obtained for the optimal configuration at time t in multistate systems. Subsequently, $I^j_{i,l}$, $I^{j^{II}}_{i,l}$, $I^D_{i,l}$, and $I^{\mathrm{D^{II}}}_{i,l}$ for the optimal structure of the consecutive-k-out-of-n: F system in multistate systems can be obtained, respectively

$$
\begin{aligned}
I^j_{i_v,l_k} &= \sum_{l=1,l\neq i}^{n} \lambda_i(v)\lambda_l(k)R_i(v)R_l(k)R^m_{opt}\left(I^j_{i_v,l_k}\right), \\
I^{j^{II}}_{i_v,l_k} &= \lambda_i(v)\lambda_l(k)R_i(v)R_l(k)R^m_{opt}\left(I^j_{i_v,l_k}\right), \\
I^D_{i_v,l_k} &= \frac{\Delta R^{\mathrm{II}}_i}{\Delta R^{\mathrm{II}}} \\
&= \frac{R^m_{opt}\left(I^B_{i,v}\right)\Delta R_i + \sum_{l=1,l\neq i}^{n} R^m_{opt}\left(I^j_{i_v,l_k}\right)\Delta R_i \Delta R_l}{\sum_{i=1}^{n} R^m_{opt}\left(I^B_{i,v}\right)\Delta R_i + \sum_{i=1}^{n}\sum_{l>i=1}^{n} R^m_{opt}\left(I^j_{i_v,l_k}\right)\Delta R_i \Delta R_l}, \\
I^{\mathrm{D}^{II}}_{i,l} &= \frac{R^m_{opt}\left(I^B_{i,v}\right)\Delta R_i + R^m_{opt}\left(I^B_{l,k}\right)\Delta R_l + R^m_{opt}\left(I^j_{i_v,l_k}\right)\Delta R_i \Delta R_l}{\sum_{i=1}^{n} R^m_{opt}\left(I^B_{i,v}\right)\Delta R_i + \sum_{i=1}^{n}\sum_{l>i=1}^{n} R^m_{opt}\left(I^j_{i_v,l_k}\right)\Delta R_i \Delta R_l}.
\end{aligned}
\tag{2.141}
$$

2.4.5 *Properties of Joint Importance Measures for Optimal Structure*

We will then give some properties of joint importance measures for the optimal structure as follows.

Theorem 2.9 *For the optimal structure of a linear consecutive-k-out-of-n: F system, assuming that when all components are independently and identically distributed, and* $p_1(t) = p_2(t) = \cdots = p_n(t) = p(t)$, *we obtain*

- *If* $\lambda_i(t) \leq \lambda_{i+1}(t)$, *then* $I^{jim^{II}}_{i,l} < I^{jim^{II}}_{i+1,l}$, *for* $i < k$;

- *If* $\lambda_l(t) \leq \lambda_{l+1}(t)$*, then* $I^{jim^{II}_{i,l}} < I^{jim^{II}_{i,l+1}}$*, for* $l < k$*;*
- *If* $\lambda_i(t) \leq \lambda_{i+1}(t)$*,* $\lambda_l(t) \leq \lambda_{l+1}(t)$*, then* $I^{jim^{II}}_{i,l} < I^{jim^{II}}_{i+1,l+1}$*, for* $i < l < k$*;*
- *For* $n - k < i < l < k$*,*

$$\begin{aligned} R_{opt}(I^j_{i,l}) &= R_{opt}\left(I^j_{i+1,l}\right) \\ &= R_{opt}\left(I^j_{i,l+1}\right) = R_{opt}\left(I^j_{i+1,l+1}\right) \end{aligned} \tag{2.142}$$

Similarly, we can draw the following inference.

Theorem 2.10 *For the optimal structure of a linear consecutive-k-out-of-n: F system, assuming that* $p_1(t) = p_2(t) = \cdots = p_n(t) = p(t)$*, we obtain*

- *For* $n - k < i < k$*,* $R_{opt}\left(I^B_i\right) = R_{opt}\left(I^B_{i+1}\right)$*;*
- *If* $\Delta R_i \leq \Delta R_{i+1}$ then $\mathrm{I}^{D^{II}}_{\mathrm{i,l}} < \mathrm{I}^{D^{II}}_{\mathrm{i+1,l}}$*, for* $i < k$*;*
- *If* $\Delta R_l \leq \Delta R_{l+1}$ then $\mathrm{I}^{D^{II}}_{\mathrm{i,l}} < \mathrm{I}^{D^{II}}_{\mathrm{i,l+1}}$*, for* $l < k$*;*
- *If* $\Delta R_i \leq \Delta R_{i+1}$ *and* $\Delta R_l \leq \Delta R_{l+1}$*, then* $I^{D^{II}}_{i,l} < I^{D^{II}}_{i+1,l+1}$*, for* $i < l < k$*.*

According to Theorems 2.9 and 2.10, when the failure rate function of component i is smaller than that of its adjacent component $i + 1$, the effect of changing component i on system reliability is less than that of changing component $i + 1$. This condition also suggests that when considering the impact of one component on the system reliability and another component fails, the component that has the greatest impact on system reliability can be selected for preventive maintenance. In addition, components with the greater importance should be improved first to obtain a significant increase in system reliability.

$I^{\mathrm{jim^{II}}}_{i,l}$ and $I^{\mathrm{D^{II}}}_{i,l}$ are the influence of component i on the system reliability when considering the sum of all other system components l. Consequently, it is not necessary to analyze the specific component l, and there is no need to discuss the impact of the change of the component l, i.e., $I^{\mathrm{jim^{II}}}_{i,l+1}$ and $I^{\mathrm{D^{II}}}_{i,l+1}$ are not discussed. Moreover, we propose Theorem 2.11 for the change of component i.

Theorem 2.11 *For the optimal structure of a linear consecutive-k-out-of-n: F system, assuming that* $p_1(t) = p_2(t) = \cdots = p_n(t) = p(t)$*, we obtain*

- *If* $\lambda_i(t) \leq \lambda_{i+1}(t)$*, then* $I^{jim^{II}}_{i,l+1} < I^{jim^{II}}_{i+1,l}$*, for* $i < k$*;*
- *If* $\Delta R_i \leq \Delta R_{i+1}$*, then* $I^{D^{II}}_{i,l+1} < I^{D^{II}}_{i+1,l}$ *for* $i < k$*.*

According to Theorem 2.11, for component i, the effect of component i on system reliability is less than that of component $i + 1$ when the influence of the sum of all other system components l is considered and the failure rate function of component i is smaller than that of component $i + 1$. Then, when the impact of component i on system reliability is considered, the most important components can be determined.

2.5 Summary

This chapter analysed system reliability improvement based on importance measures. In order to obtain the maximal improvement of system reliability, it analysed the direction of the fastest improvement in system performance by investigating the gradient calculation and geometric significance of important measures. Meanwhile, the chapter proposed a multi-criteria importance measure and its applications to reliability improvement. Then, importance measures for system reconfiguration was studied. The chapter also considered the component reliability importance measures with respect to the changes of the optimal component sequencing. In addition, the joint reliability importance (JRI) analyses how much influence the interaction among components has on the reliability of the k-out-of-n: F system and the consecutive k-out-of-n: F system. Further, this chapter proposed joint importance measures for system reconfiguration. The relationships between component reliability and joint importance measures were then discussed.

References

1. Banerjee S, Mondal S, Chatterjee P, Pramanick AK (2021) An intercriteria correlation model for sustainable automotive body material selection. J Ind Eng Decis Making 2(1):8–14
2. Boddu P, Xing L (2012) Redundancy allocation for k-out-of-n: G systems with mixed spare types. In: 2012 proceedings annual reliability and maintainability symposium, IEEE, pp 1–6
3. Boddu P, Xing L (2013) Reliability evaluation and optimization of series-parallel systems with k-out-of-n: G subsystems and mixed redundancy types. Proc Inst Mech Eng Part O: J Risk Reliab 227(2):187–198
4. Borgonovo E, Apostolakis GE (2001) A new importance measure for risk-informed decision making. Reliab Eng Syst Saf 72(2):193–212
5. Cai B, Fan H, Shao X, Liu Y, Liu G, Liu Z, Ji R (2021) Remaining useful life re-prediction methodology based on wiener process: Subsea Christmas tree system as a case study. Comput Ind Eng 151:106983
6. Chang G, In-Hang C, Cui L, Hwang F (2000) Reliabilities of consecutive-k systems, vol 4. Springer
7. Chang GJ, Cui L, Hwang FK (1999) New comparisons in Birnbaum importance for the consecutive-k-out-of-n system. Probab Eng Informational Sci 13(2):187–192
8. Cui L, Hawkes AG (2008) A note on the proof for the optimal consecutive-k-out-of-n: G line for $n \leq 2k$. J Stat Plann Infer 138(5):1516–1520
9. Do P, Bérenguer C (2020) Conditional reliability-based importance measures. Reliab Eng Syst Saf 193:106633
10. Du Y, Si S, Cai Z, Jin T (2019) Bayesian importance measures for network edges under saturated Lagrangian Poisson failures. IEEE Trans Reliab 70(1):110–120
11. Dui H, Si S, Wu S, Yam RC (2017) An importance measure for multistate systems with external factors. Reliab Eng Syst Saf 167:49–57
12. Dui H, Si S, Yam RC (2018) Importance measures for optimal structure in linear consecutive-k-out-of-n systems. Reliab Eng Syst Saf 169:339–350
13. Eryilmaz S (2013) Joint reliability importance in linear m-consecutive-k-out-of-n: F systems. IEEE Trans Reliab 62(4):862–869
14. Eryilmaz S, Coolen FP, Coolen-Maturi T (2018) Marginal and joint reliability importance based on survival signature. Reliab Eng Syst Saf 172:118–128

15. Gao X, Cui L, Li J (2007) Analysis for joint importance of components in a coherent system. Eur J Oper Res 182(1):282–299
16. Hong JS, Koo HY, Lie CH (2002) Joint reliability importance of k-out-of-n systems. Eur J Oper Res 142(3):539–547
17. Hwang FK, Cui L, Chang JC, Lin WD (2000) Comments on "reliability and component importance of a consecutive-k-out-of-n system" by Zuo. Microelectron Reliab 40(6):1061–1063
18. Jalali A, Hawkes A, Cui L, Hwang F (2005) The optimal consecutive-k-out-of-n: G line for $n \leq 2k$. J Stat Plan Infer 128(1):281–287
19. Kuo W, Zhang W, Zuo M (1990) A consecutive-k-out-of-n: G system: the mirror image of a consecutive-k-out-of-n: F system. IEEE Trans Reliab 39(2):244–253
20. Levitin G (2002) Optimal allocation of elements in a linear multi-state sliding window system. Reliab Eng Syst Saf 76(3):245–254
21. Levitin G (2003) Optimal allocation of multistate elements in a linear consecutively-connected system. IEEE Trans Reliab 52(2):192–199
22. Levitin G, Ben-Haim H (2008) Importance of protections against intentional attacks. Reliab Eng Syst Saf 93(4):639–646
23. Levitin G, Lisnianski A (2001) Reliability optimization for weighted voting system. Reliab Eng Syst Saf 71(2):131–138
24. Levitin G, Xing L, Dai Y (2013) Sequencing optimization in k-out-of-n cold-standby systems considering mission cost. Int J Gen Syst 42(8):870–882
25. Lin FH, Kuo W (2002) Reliability importance and invariant optimal allocation. J Heuristics 8:155–171
26. Lin FH, Kuo W, Hwang F (1999) Structure importance of consecutive-k-out-of-n systems. Oper Res Lett 25(2):101–107
27. Liu M, Wang D, Zhao J, Si S (2022) Importance measure construction and solving algorithm oriented to the cost-constrained reliability optimization model. Reliab Eng Syst Saf 222:108406
28. Mahmoud B, Eryilmaz S (2014) Joint reliability importance in a binary k-out-of-n: G system with exchangeable dependent components. Qual Technol Quant Manage 11(4):453–460
29. Malon DM (1984) Optimal consecutive-2-out-of-n: F component sequencing. IEEE Trans Reliab 33(5):414–418
30. Malon DM (1985) Optimal consecutive-k-out-of-n: F component sequencing. IEEE Trans Reliab 34(1):46–49
31. Mo Y, Xing L, Cui L, Si S (2017) Mdd-based performability analysis of multi-state linear consecutive-k-out-of-n: F systems. Reliab Eng Syst Saf 166:124–131
32. Papastavridis S (1987) The most important component in a consecutive-k-out-of-n: F system. IEEE Trans Reliab 36(2):266–268
33. Ramirez-Marquez JE, Coit DW (2005) Composite importance measures for multi-state systems with multi-state components. IEEE Trans Reliab 54(3):517–529
34. Ramirez-Marquez JE, Coit DW (2007) Multi-state component criticality analysis for reliability improvement in multi-state systems. Reliab Eng Syst Saf 92(12):1608–1619
35. Rani M, Jain K, Dewan I (2011) On conditional marginal and conditional joint reliability importance. Int J Reliab Qual Saf Eng 18(02):119–138
36. Si S, Cai Z, Sun S, Zhang S (2010) Integrated importance measures of multi-state systems under uncertainty. Comput Ind Eng 59(4):921–928
37. Si S, Levitin G, Dui H, Sun S (2013) Component state-based integrated importance measure for multi-state systems. Reliab Eng Syst Saf 116:75–83
38. Steffen T, Schiller F, Blum M, Dixon R (2013) Analysing the reliability of actuation elements in series and parallel configurations for high-redundancy actuation. Int J Syst Sci 44(8):1504–1521
39. Tong Y (1985) A rearrangement inequality for the longest run, with an application to network reliability. J Appl Probab 22(2):386–393
40. Wang Y, Bi L, Lin S, Li M, Shi H (2017) A complex network-based importance measure for mechatronics systems. Physica A: Stat Mech Appl 466:180–198

41. Yam RC, Zuo MJ, Zhang YL (2003) A method for evaluation of reliability indices for repairable circular consecutive-k-out-of-n: F systems. Reliab Eng Syst Saf 79(1):1–9
42. Yao Q, Zhu X, Kuo W (2011) Heuristics for component assignment problems based on the Birnbaum importance. IIE Trans 43(9):633–646
43. Yun WY, Kim GR, Yamamoto H (2012) Economic design of a load-sharing consecutive k-out-of-n: F system. IIE Trans 44(1):55–67
44. Zhu X, Yao Q, Kuo W (2012) Patterns of the Birnbaum importance in linear consecutive-k-out-of-n systems. IIE Trans 44(4):277–290
45. Zhu X, Fu Y, Yuan T, Wu X (2017) Birnbaum importance based heuristics for multi-type component assignment problems. Reliab Eng Syst Saf 165:209–221
46. Zio E, Podofillini L (2003) Monte Carlo simulation analysis of the effects of different system performance levels on the importance of multi-state components. Reliab Eng Syst Saf 82(1):63–73
47. Zio E, Podofillini L (2006) Accounting for components interactions in the differential importance measure. Reliab Eng Syst Saf 91(10–11):1163–1174
48. Zuo M, Kuo W (1990) Design and performance analysis of consecutive-k-out-of-n structure. Naval Res Logist 37(2):203–230

Chapter 3
Importance Measures for Optimisation of Cost Independent Maintenance Policies

Abstract The introduction of reliability importance measures is rooted their applications in the improvement of engineered systems. This chapter aims at investigating the applications of reliability importance measures in maintenance policy optimisation by considering various resource constrains including budget constraints and the size of the maintenance team.

Keywords Preventive maintenance · Maintenance policy · Maintenance priority

Maintenance policies are usually developed to provide guidance on maintenance for failed components. Due to the differences and interdependencies between components, different maintenance policies, which mainly specify maintenance intervals and the effectiveness of maintenance, can have different impacts on the reliability/availability of a system. In this chapter, optimal cost-based maintenance policies are investigated based on importance measures. Performance-based importance measures, reliability-based importance measures, and importance-based optimal maintenance policies are proposed, respectively. This chapter mainly proposes an extended joint integrated importance measure (JIIM) to guide the selection of components for PM with the aim to maximise system performance. Besides, based on the extension to integrated importance and system resilience, the chapter proposes two importance measures, which define component maintenance priorities, considering maintenance priorities for constrained resources. The chapter finally proposes a new importance measure for maintenance policy optimisation.

© The Author(s), under exclusive license to Springer Nature Switzerland AG 2024

H. Dui and S. Wu, *Importance-Informed Reliability Engineering*, Springer Series in Reliability Engineering, https://doi.org/10.1007/978-3-031-52455-4_3

3.1 Performance-Based Importance Measures for Optimisation of Maintenance Policies

For a multi-state system, the expectation of system performance can be defined by

$$U(\boldsymbol{X}(t)) = \sum_{j=1}^{M} a_j \Pr[\Phi(\boldsymbol{X}(t)) = j]$$
$$= \sum_{j=1}^{M} a_j \Pr[\Phi(X_1(t), X_2(t), \ldots, X_n(t)) = j], \tag{3.1}$$

where a_j is the system performance level while the system is at state j, $\Pr[\Phi(\boldsymbol{X}(t)) = j]$ is the probability that the system is at state j at time t. The system structure function $\Phi(\boldsymbol{X}(t))$ is related to states of all components, and a component has only two states (working and failed). Thus, $\Pr[\Phi(\boldsymbol{X}(t)) = j]$ is a function of the probabilities of components sojourning at states, i.e., $\Pr[\Phi(\boldsymbol{X}(t)) = j] = f_j(R_1(t), R_2(t), \ldots, R_n(t))$.

Based on the loss of system performance, the integrated importance of component i is defined as

$$I_i^{im}(t) = \sum_{j=1}^{M} a_j R_i(t)\lambda_i(t)\{\Pr[\Phi(1_i, \boldsymbol{X}(t)) = j] - \Pr[\Phi(0_i, \boldsymbol{X}(t)) = j]\}. \tag{3.2}$$

Proposition 3.1 *The change of system performance in a unit of time is the sum of integrated importance values of all system components, i.e.,* $\frac{dU(\boldsymbol{X}(t))}{dt} = -\sum_{i=1}^{n} I_i^{im}(t)$.

Proof Based on Eq. (3.1), we have

$$\frac{dU(\boldsymbol{X}(t))}{dt} = \frac{d}{dt}\sum_{j=1}^{M} a_j f_j(R_1(t), R_2(t), \ldots, R_n(t))$$
$$= \sum_{j=1}^{M} a_j \sum_{i=1}^{n} \frac{dR_i(t)}{dt} \frac{\partial f_j(R_1(t), R_2(t), \ldots, R_n(t))}{\partial R_i(t)}$$
$$= \sum_{j=1}^{M} a_j \sum_{i=1}^{n} \frac{dR_i(t)}{dt} \frac{\partial \Pr[\Phi(\boldsymbol{X}(t)) = j]}{\partial R_i(t)}$$
$$= \sum_{j=1}^{M}\sum_{i=1}^{n} a_j \frac{dR_i(t)}{dt} \frac{\partial \Pr[\Phi(\boldsymbol{X}(t)) = j]}{\partial R_i(t)}.$$

Because

$$\Pr[\Phi(\boldsymbol{X}(t)) = j] = \Pr[X_i(t) = 1]\Pr[\Phi(1_i, \boldsymbol{X}(t)) = j] + \Pr[X_i(t) = 0]\Pr[\Phi(0_i, \boldsymbol{X}(t)) = j] \quad (3.3)$$

and $\lambda_i(t) = -\frac{dR_i(t)/dt}{R_i(t)}$, then we have

$$\frac{dU(\boldsymbol{X}(t))}{dt} = -\sum_{i=1}^{n}\sum_{j=1}^{M} a_j R_i(t)\lambda_i(t)\{\Pr[\Phi(1_i, \boldsymbol{X}(t)) = j] - \Pr[\Phi(0_i, \boldsymbol{X}(t)) = j]\}. \quad (3.4)$$

Substituting Eq. (3.2) into Eq. (3.4), we can obtain

$$\frac{dU(\boldsymbol{X}(t))}{dt} = -\sum_{i=1}^{n} I_i^{im}(t).$$

Because the system performance $U(\boldsymbol{X}(t))$ is a decreasing function of time t, $\frac{dU(\boldsymbol{X}(t))}{dt} < 0$. Then the change of system performance in a unit of time is the sum of integrated importance values of all system components. □

From Proposition 3.1, $I_i^{im}(t)$ can be interpreted as the contribution of the change of component i's state to the change of the system performance per unit of time.

3.1.1 An Extended Joint Integrated Importance Measure

The improvement of the system performance made by repairing component m is

$$U(1_m, \boldsymbol{X}(t)) - U(0_m, \boldsymbol{X}(t)) = \sum_{j=1}^{M} a_j\{\Pr[\Phi(1_m, \boldsymbol{X}(t)) = j] - \Pr[\Phi(0_m, \boldsymbol{X}(t)) = j]\}. \quad (3.5)$$

When component m fails, the system performance $U(\boldsymbol{X}(t))$ becomes $U(0_m, \boldsymbol{X}(t))$. Similar to Eq. (3.2), we have

$$\frac{dU(0_m, \boldsymbol{X}(t))}{dt} = -\sum_{i=1, i\neq m}^{n}\sum_{j=1}^{M} a_j R_i(t)\lambda_i(t)\{\Pr[\Phi(0_m, 1_i, \boldsymbol{X}(t)) = j] - \Pr[\Phi(0_m, 0_i, \boldsymbol{X}(t)) = j]\}. \quad (3.6)$$

Let

$$I_i^{im}(t)_{X_m(t)=0}$$
$$=\sum_{j=1}^{M} a_j R_i(t)\lambda_i(t)\{\Pr[\Phi(0_m, 1_i, \boldsymbol{X}(t)) = j] - \Pr[\Phi(0_m, 0_i, \boldsymbol{X}(t)) = j]\}. \quad (3.7)$$

Then, $I_i^{im}(t)_{X_m(t)=0}$ represents the contribution of component i to the change of system performance in a unit of time after component m is failed. But if component m is a critical component (if component m fails, then the system also fails, also referred to as a single-point of failure), $U(0_m, \boldsymbol{X}(t)) = 0$. In this case $I_i^{im}(t)_{X_m(t)=0}$ cannot be used to rank other system components because the values $I_i^{im}(t)_{X_m(t)=0}$ are all zero.

If component m works, the system performance $U(\boldsymbol{X}(t))$ becomes $U(1_m, \boldsymbol{X}(t))$. Similar to Eq. (3.2), we have

$$\frac{dU(1_m, \boldsymbol{X}(t))}{dt} = -\sum_{i=1, i\neq m}^{n}\sum_{j=1}^{M} a_j R_i(t)\lambda_i(t)\Pr[\Phi(1_m, 1_i, \boldsymbol{X}(t)) = j]$$
$$-\sum_{i=1, i\neq m}^{n}\sum_{j=1}^{M} a_j R_i(t)\lambda_i(t)\Pr[\Phi(1_m, 0_i, \boldsymbol{X}(t)) = j]. \quad (3.8)$$

Let

$$I_i^{im}(t)_{X_m(t)=1}$$
$$=\sum_{j=1}^{M} a_j R_i(t)\lambda_i(t)\{\Pr[\Phi(1_m, 1_i, \boldsymbol{X}(t)) = j] - \Pr[\Phi(1_m, 0_i, \boldsymbol{X}(t)) = j]\}. \quad (3.9)$$

Then $I_i^{im}(t)_{X_m(t)=1}$ represents the contribution of component i to the change of system performance in a unit of time when component m works.

Definition 3.1 If component m is failed and being repaired, the JIM between components i and m is defined as $I_i^{im}(t)_{X_m(t)} = I_i^{im}(t)_{X_m(t)=1} - I_i^{im}(t)_{X_m(t)=0}$.

Proposition 3.2 *If component m fails, the component, component i for example, with the maximum $I_i^{im}(t)_{X_m(t)}$ should be selected for PM (Preventive Maintenance) so that the system's performance has the largest improvement.*

Proof When component m is repaired, the system performance improvement in each unit of time can be obtained by

$$\frac{d\left(U(0_m, \boldsymbol{X}(t)) - U(1_m, \boldsymbol{X}(t))\right)}{dt} = \frac{dU(0_m, \boldsymbol{X}(t))}{dt} - \frac{dU(1_m, \boldsymbol{X}(t))}{dt}. \quad (3.10)$$

Substituting Eq. (3.7) into Eq. (3.6),

$$\frac{dU(0_m, \boldsymbol{X}(t))}{dt} = -\sum_{i=1, i\neq m}^{n} I_i^{im}(t)_{X_m(t)=0}. \tag{3.11}$$

Similarly, substituting Eq. (3.9) into Eq. (3.8),

$$\frac{dU(1_m, \boldsymbol{X}(t))}{dt} = -\sum_{i=1, i\neq m}^{n} I_i^{im}(t)_{X_m(t)=1}. \tag{3.12}$$

Then,

$$\begin{aligned}\frac{dU(0_m, \boldsymbol{X}(t))}{dt} - \frac{dU(1_m, \boldsymbol{X}(t))}{dt} &= \sum_{i=1, i\neq m}^{n} (I_i^{im}(t)_{X_m(t)=1} - I_i^{im}(t)_{X_m(t)=0}) \\ &= \sum_{i=1, i\neq m}^{n} I_i^{im}(t)_{X_m(t)}. \end{aligned} \tag{3.13}$$

Thus, while component m is being repaired, the improvement of system performance in a unit of time is the sum of JIIM values of all other system components, in which $I_i^{im}(t)_{X_m(t)}$ is the contribution of component i to the changes of the system performance in a unit of time while component m is being repaired. Therefore, the maximum $I_i^{im}(t)_{X_m(t)}$ could lead to the maximal improvement of the system performance. □

In many cases, maintenance resource may support more than one component to be maintained. From Proposition 3.2, we should select the components for PM following the ranking of the component JIIMs.

In fact,

$$\begin{aligned} I_i^{im}(t)_{X_m(t)} &= I_i^{im}(t)_{X_m(t)=1} - I_i^{im}(t)_{X_m(t)=0} \\ &= \sum_{j=1}^{M} a_j R_i(t)\lambda_i(t)\{\Pr[\Phi(1_m, 1_i, \boldsymbol{X}(t)) = j] \\ &\quad + \Pr[\Phi(0_m, 0_i, \boldsymbol{X}(t)) = j] \\ &\quad - \Pr[\Phi(1_m, 0_i, \boldsymbol{X}(t)) = j] \\ &\quad - \Pr[\Phi(0_m, 1_i, \boldsymbol{X}(t)) = j]\}. \end{aligned} \tag{3.14}$$

Equation (3.14) can be rewritten as

$$I_i^{im}(t)_{X_m(t)} = \sum_{j=1}^{M} a_j R_i(t)\lambda_i(t) I_{i,m}^{j}(t), \tag{3.15}$$

where $I_{i,m}^{j}(t)$ is the joint reliability importance between components i and m at time t. Based on the proposed JIIM, we build the matrix shown in Eq. (3.16), where we assume $I_i^{im}(t)_{X_i(t)} = 0$.

$$A = \begin{bmatrix} 0 & I_2^{im}(t)_{X_1(t)} & \cdots & I_{n-1}^{im}(t)_{X_1(t)} & I_n^{im}(t)_{X_1(t)} \\ I_1^{im}(t)_{X_2(t)} & 0 & \cdots & I_{n-1}^{im}(t)_{X_2(t)} & I_n^{im}(t)_{X_2(t)} \\ \vdots & \vdots & \ddots & \vdots & \vdots \\ I_1^{im}(t)_{X_m(t)} & I_2^{im}(t)_{X_m(t)} & \cdots & I_{n-1}^{im}(t)_{X_m(t)} & 0 \end{bmatrix}. \tag{3.16}$$

The mth row in the matrix of Eq. (3.16) represents the contribution of each system component (other than component m) to the improvement of the system performance in a unit of time when the failed component m is maintained. In general, $I_i^{im}(t)_{X_m(t)}$ and $I_m^{im}(t)_{X_i(t)}$ are different.

Proposition 3.3 *If $R_i(t) = R_m(t)$, then $I_i^{im}(t)_{X_m(t)} = I_m^{im}(t)_{X_i(t)}$ for $i \neq m$.*

Proof It can be verified directly from the fact that

$$I_i^{im}(t)_{X_m(t)} = \sum_{j=1}^{M} a_j R_i(t)\lambda_i(t) I_{i,m}^{j}(t) \tag{3.17}$$

and $I_{i,m}^{j}(t) = I_{m,i}^{j}(t)$. □

From Proposition 3.3, if $R_1(t) = R_2(t) = K = R_n(t)$, then Matrix A is symmetrical. Moreover, the relationship between the proposed JIIM and the IIM is presented in the following proposition.

Proposition 3.4 $I_i^{im}(t) = R_m(t) I_i^{im}(t)_{X_m(t)} + I_i^{im}(t)_{X_m(t)=0}$.

Proof Based on Eq. (3.2), we have

$$\begin{aligned} I_i^{im}(t) &= \sum_{j=1}^{M} a_j R_i(t)\lambda_i(t)\{\Pr[\Phi(1_i, \boldsymbol{X}(t)) = j] - \Pr[\Phi(0_i, \boldsymbol{X}(t)) = j]\} \\ &= \sum_{j=1}^{M} a_j R_i(t)\lambda_i(t) R_m(t) \Pr[\Phi(1_m, 1_i, \boldsymbol{X}(t)) = j] \\ &\quad + (1 - R_m(t)) \Pr[\Phi(0_m, 1_i, \boldsymbol{X}(t)) = j] - R_m(t) \Pr[\Phi(1_m, 0_i, \boldsymbol{X}(t)) = j] \\ &\quad + (1 - R_m(t)) \Pr[\Phi(0_m, 0_i, \boldsymbol{X}(t)) = j] \\ &= R_m(t) \sum_{j=1}^{M} a_j R_i(t)\lambda_i(t)\{\Pr[\Phi(1_m, 1_i, \boldsymbol{X}(t)) = j] - \Pr[\Phi(1_m, 0_i, \boldsymbol{X}(t)) = j]\} \\ &\quad + (1 - R_m(t)) \sum_{j=1}^{M} a_j R_i(t)\lambda_i(t)\{\Pr[\Phi(0_m, 1_i, \boldsymbol{X}(t)) = j] - \Pr[\Phi(0_m, 0_i, \boldsymbol{X}(t)) = j]\}. \end{aligned} \tag{3.18}$$

Substituting Eqs. (3.7) and (3.9) into Eq. (3.18), we have

$$\begin{aligned} I_i^{im}(t) &= R_m(t) I_i^{im}(t)_{X_m(t)=1} + (1 - R_m(t)) I_i^{im}(t)_{X_m(t)=0} \\ &= R_m(t)(I_i^{im}(t)_{X_m(t)=1} - I_i^{im}(t)_{X_m(t)=0}) + I_i^{im}(t)_{X_m(t)=0}. \end{aligned} \tag{3.19}$$

Finally, according to the JIIM definition, we obtain

$$I_i^{im}(t) = R_m(t) I_i^{im}(t)_{X_m(t)} + I_i^{im}(t)_{X_m(t)=0}. \tag{3.20}$$

From Proposition 3.4, if component m is a critical component, then $I_i^{im}(t)_{X_m(t)=0} = 0$. Hence $I_i^{im}(t) = R_m(t) I_i^{im}(t)_{X_m(t)}$. This shows that JIIM coincides with IIM when guiding the selection of the critical component m in PM.

While component m is being repaired, and given the fixed maintenance cost C, we should select components for PM to maximise the expected system performance. In other words, we need to solve the following integer programming problem.

$$\max_{z_i} \sum_{i \neq m} z_i I_i^{im}(t)_{X_{m(t)}}, \tag{3.21}$$

subject to

$$\sum_{i \neq m} c_i z_i \leq C$$

and

$$z_i \in \{0, 1\},$$

where c_i represents the maintenance cost for component i and z_i is the decision variable representing whether component i should be maintained or not. Note that z_i can only take values from 0 and 1. To derive the optimal maintenance policy, it suffices to test the 2^{n-1} combinations of z_i. Under the optimal maintenance policy $\{z_i^*, i \neq m\}$, the number of maintained components is $\sum_{i \neq m} z_i^*$. Note that when each component has the same maintenance cost, the components for PM can be determined following the ranking of components JIIM. However, considering the cost of the PM on heterogeneous components, components with larger JIIM may also incur larger PM cost. In this case, to allocate PM priority to the component with large JIIM is not always an optimal solution. Thus, we should use the above integer programming Eq. (3.21) to determine the components for PM.

If failed component m is a critical component (or a single-point of failure), the system fails and stops working. In this case PM can be performed on all the other system components. Based on the limited budget or maintenance resource, JIIM $I_i^{im}(t)_{X_m(t)}$ can be used to choose the most important ones for PM. When the failed component m is a non-critical component, the system does not fail. In order to not interrupt the system, PM must be performed on only non-critical components, i.e., components satisfying $\Phi(0_m, 0_i, \boldsymbol{X}(t)) \neq 0$. Again, based on limited maintenance resource, JIIM $I_i^{im}(t)_{X_m(t)}$ can be used to choose the most important non-critical components for PM.

3.1.2 Two Importance Measures

3.1.2.1 Resilience Efficiency Importance Measure

When N' components fail simultaneously, the system performance will degrade from $U(\boldsymbol{X}(t))$ to $U(S_{N'}, \boldsymbol{X}(t))$. In this process, the rate of change in system performance is expressed as

$$\begin{aligned}\frac{dU\left(S_{N'}, \boldsymbol{X}(t)\right)}{dt} &= -\sum_{i=1}^{n}\sum_{j=1}^{M} a_j R_i(t)\lambda_i(t)\left\{\Pr\left[\Phi(S_{N'}, B_i, \boldsymbol{X}(t)) = j\right]\right.\\ &\quad \left. - \sum_{\beta=1}^{B-1} k_\beta^i \Pr\left[\Phi(S_{N'}, \beta_i, \boldsymbol{X}(t)) = j\right]\right\}\\ &= -\sum_{i=1, i\notin N'}^{n} I_i^{\mathrm{im}}(t)_{S_{N'}},\end{aligned} \tag{3.22}$$

where N' is the set of failed components, in which there is N' failed components; $S_{N'}$ is the set of the failure states of failed components in N', element β_m in which represents the failure state of failed component m, and $m \in N'$.

$I_i^{im}(t)_{S_{N'}}$ represents the contribution of component i to the change of system performance in a unit of time while components in N' fail. For component $m \in N'$, it is obvious that it has no contribution to system performance change. Thus, we define $I_m^{im}(t)_{S_{N'}} = 0$, where $m \in N'$.

Further, the loss of system performance in a unit of time while a failure occurs is represented as

$$\begin{aligned}\frac{d\left(U(\boldsymbol{X}(t)) - U\left(S_{N'}, \boldsymbol{X}(t)\right)\right)}{dt} &= \frac{dU(\boldsymbol{X}(t))}{dt} - \frac{dU\left(S_{N'}, \boldsymbol{X}(t)\right)}{dt}\\ &= -\sum_{i=1}^{n} I_i^{im}(t) + \sum_{i=1}^{n} I_i^{im}(t)_{S_{N'}}\\ &= -\sum_{i=1}^{n}\left(I_i^{im}(t) - I_i^{im}(t)_{S_{N'}}\right).\end{aligned} \tag{3.23}$$

We define the vulnerability importance measure (VIM) of component i corresponding to failure event $S_{N'}$ as

$$I_i^{v}(t)_{S_{N'}} = I_i^{im}(t) - I_i^{im}(t)_{S_{N'}}. \tag{3.24}$$

According to Eq. (3.23), the loss of system performance in a unit of time while a failure occurs is equal to the sum of the VIM values of all components in a system. $I_i^v(t)_{S_{N'}}$ represents the contribution of component i to the change of system performance in a unit of time during the failure event $S_{N'}$. For $m \in N'$, $I_m^{im}(t)_{S_{N'}(t)} = 0$. Hence, $I_m^v(t)_{S_{N'}} = I_m^{im}(t)$, $m \in N'$, i.e., the contribution of component m to the change of system performance in the course of its own failure is equal to that while no failure occurs in a system.

For a resilient system, system performance can be recovered gradually by adopting a maintenance policy to repair the failed components. While failed components are being repaired, the improvement of system performance is $U(B_{R'}, S_{N' \backslash R'}, \boldsymbol{X}(t)) - U(S_{N'}, \boldsymbol{X}(t))$. The rate of change of system performance in the process of repair can be represented as

$$\begin{aligned}
&\frac{d\left(U\left(B_{R'}, S_{N' \backslash R'}, \boldsymbol{X}(t)\right) - U\left(S_{N'}, \boldsymbol{X}(t)\right)\right)}{dt} \\
&= \frac{dU\left(B_{R'}, S_{N' \backslash R'}, \boldsymbol{X}(t)\right)}{dt} - \frac{dU\left(S_{N'}, \boldsymbol{X}(t)\right)}{dt} \\
&= -\sum_{i=1, i \notin N'}^{n} I_i^{im}(t)_{X_{R'}(t)=B, S_{N' \backslash R'}} + \sum_{i=1, i \notin N'}^{n} I_i^{im}(t)_{S_{N'}} \\
&= -\sum_{i=1, i \notin N'}^{n} \left(I_i^{im}(t)_{X_{R'}(t)=B, S_{N' \backslash R'}} - I_i^{im}(t)_{S_{N'}}\right),
\end{aligned} \tag{3.25}$$

where $R' \subseteq N'$; $S_{N' \backslash R'}$ is the set of the states of failed components that have not been repaired in N', $S_{N' \backslash R'} \subseteq S_{N'}$; $B_{R'}$ represents the set of the states of components in R', in which all components are in the perfect state.

Definition 3.2 The recovery importance measure (RIM) of component i in the course of component set R being repaired is given by

$$I_i^r(t)_{S_{N'}}^{X_{R'}(t)} = I_i^{\text{im}}(t)_{X_R(t)=B, S_{N' \backslash R}} - I_i^{\text{im}}(t)_{S_{N'}}. \tag{3.26}$$

According to Eq. (3.25), the improvement of system performance in a unit of time in the course of component set R' being repaired is equal to the sum of the RIM values of all non-failure components in a system, in which $I_i^r(t)_{S_{N'}}^{X_{R'}(t)}$ represents the contribution of component i to system performance improvement in a unit of time in the course of component set R' being repaired after failure event $S_{N'}$. When only component m fails, $I_i^r(t)_{\beta_m}^{X_m(t)} = I_i^{im}(t)_{X_m(t)=B} - I_i^{im}(t)_{X_m(t)=\beta} = I_i^{im}(t)_{X_m(t)}$, indicating that in the course of the only failed component m being repaired, RIM is consistent with JIIM.

In order to maximise the recovery efficiency of system performance, priority should be given to the component with the best repair benefit during maintenance. In terms of system resilience, the repair efficiency cannot be measured by the absolute value of system performance recovery caused by the repair of components alone. The loss of system performance in the process of failure should be taken into account, considering the effect of the repair of components to the recovery of system performance loss caused by failure.

Definition 3.3 After failure event $S_{N'}$, the resilience efficiency importance measure (REIM) of failed component m is defined as the degree to which the individual repair of component m recovers the loss of system performance in a unit of time, i.e.,

$$I_m^e(t)^{S_{N'}} = \frac{\dfrac{d\left(U\left(B_m, S_{N'\{m\}}, X(t)\right) - U\left(S_{N'}, X(t)\right)\right)}{dt}}{\dfrac{d\left(U(X(t)) - U\left(S_{N'}, X(t)\right)\right)}{dt}}. \tag{3.27}$$

Substituting (3.18) and (3.19) into (3.20), we have

$$I_m^e(t)_{S_{N'}} = \frac{\sum_{i=1, i \notin N'}^{n} I_i^{\mathrm{r}}(t)_{S_{N'}}^{X_m(t)}}{\sum_{i=1}^{n} I_i^{\mathrm{v}}(t)_{S_{N'}}}. \tag{3.28}$$

According to Eq. (3.28), the REIM value of component m is equal to the ratio of the sum of RIM values of all non-failure components in a system to the sum of VIM values of all components in a system. The larger the REIM value of a component, the higher the recovery efficiency of the system performance when taking the lead to repair it, which means that the component should be given a higher maintenance priority. After a failure occurs, for all failed components, the sum of VIM values of all components corresponding to them is the same. Hence, when comparing different failed components under the same failure event, using REIM to determine the maintenance priority of component m is consistent with using the absolute amount of system performance recovery in a unit of time while only the component is begin repaired.

3.1.2.2 REIMII and MEM

REIMII REIMII is an extension of REIM that evaluates the rehabilitation benefits of components under different maintenance sequences. After a failure event $S_{N'}$, if only a component can be repaired at a time, maintenance policies can be divided by maintenance sequence of failed components, and each strategy can be represented as a vector composed of failed components sorted in maintenance sequence, i.e.

$$W_p = \left(A_1, A_2, \ldots, A_h, \ldots A_{n'}\right), p = 1, 2, \ldots, A_{n'}^{n'}, \tag{3.29}$$

where A_h is the h-th component that is repaired. We use REIMII to denote the recovery system performance loss per unit of time after a failure event $S_{N'}$, defined as follows,

$$I_m^{eII}(t)^{S_{N'},W_p} = \frac{\dfrac{d\left(U\left(B_{F_{W_p}^m}, S_{N'\backslash F_{W_p}^m}, X(t)\right) - U\left(S_{N'}, X(t)\right)\right)}{dt}}{\dfrac{d\left(U(X(t)) - U\left(S_{N'}, X(t)\right)\right)}{dt}}. \tag{3.30}$$

Substituting Eqs. (3.23) and (3.25) into Eq. (3.28), we have

$$I_m^{eII}(t)^{S_{N'},W_p} = \frac{\sum_{i=1,i\notin N'}^{n} I^r(t)_{S_{N'}}^{X_{F_{W_p}^m}(t)}}{\sum_{i=1}^{n} I^v(t)_{S_{N'}}}, \tag{3.31}$$

where $F_{W_p}^m$ is the set of failed components that have been repaired by the time when component m is repaired to be perfect, according to a maintenance policy W_p, where $F_{W_p}^m \subseteq N'$.

The REIMII of failed component m reflects the recovery efficiency of system performance loss in the whole repair process from start to the time when component m is restored to perfect. As can be seen from Eqs. (3.30) and (3.31), REIMII is closely related to the integrated importance of the component itself as well as to the structural function of the system.

MEM In order to measure the effect of different maintenance policies on the recovery efficiency of system performance, the maintenance efficiency measure (MEM) is proposed as below.

$$I_{W_p}^m(t) = \frac{1}{z_p}\sum_{W_p} I_m^{eII}(t)^{S_{N'},W_p}, \tag{3.32}$$

where z_p is the number of components included in maintenance policy W_p.

The goal of maintenance policy optimisation is to find the maintenance sequence with the highest repair efficiency, so as to optimise the overall recovery efficiency of system performance in the repair process, which can be expressed by

$$\max_{W_p \in \{W_1, W_2, \ldots, W_{A_{n'}^{n'}}\}} I_{W_p}^m(t). \tag{3.33}$$

3.2 Reliability-Based Importance Measures for Optimisation of Maintenance Policies

This section makes the following assumptions.

(1) The multi-state system under discussion is monotone and coherent.
(2) The state space of component i is $\{0, 1, \ldots, M_i\}$ and the state space of the system is $\{0, 1, \ldots, M\}$, where 0 represents the complete failure of the system or components and $M_i\,(M)$ is the perfect functioning state of component i (system).
(3) All components (states) and the system (state) are statistically independent with each other.

(4) Each state of a component is characterised by a different level of performance. Precisely, the states of a component, component i say, are numbered according to decreasing performance levels, from M_i to 0.

Let $a_0 \leq a_1 \leq \ldots \leq a_M$ be the performance levels corresponding to the state space $\{0, 1, \ldots, M\}$ of a multistate system. Let $a_0 = 0$, without loss of generality, then the expected performance of the system can be defined by:

$$\begin{aligned} U(\mathbf{X}(t)) &= \sum_{v=1}^{M} a_v \Pr(\Phi(\mathbf{X}(t)) = v) \\ &= \sum_{v=1}^{M} a_v \Pr(\Phi(X_1(t), X_2(t), \ldots X_n(t)) = v)\,. \end{aligned} \quad (3.34)$$

The Griffith importance of state m of component i is defined as

$$I_{i_m}^{G}(t) = \sum_{v=1}^{M} (a_v - a_{v-1}) \left[\Pr(\Phi(m_i, X(t)) \geq v) - \Pr(\Phi((m-1)_i, X(t)) \geq v)\right]. \quad (3.35)$$

Wu et al. [4] defined the component maintenance priority (CMP) of binary systems to prioritise the components to be maintained during the repair of the failed component.

$$I_{j|i}^{M}(t) = H_{j|i} \frac{\partial \phi(\lambda_i, \mathbf{p}_i(t))}{\partial p_j(t)}, \quad (3.36)$$

where $p_j(t)$ is the reliability of component j, $(0_i, 0_j, \ldots, \mathbf{1}_{i,j})$ represents that both components i and j stop working while all of the other components are working, $\lambda_i = \chi\{\phi(1_1, 1_2, \ldots, 1_{i-1}, 0_i, 1_{i+1}, \ldots, 1_n) = 0\}$, and $\phi(\mathbf{p}_i(t))$ is the system reliability as a function of $\mathbf{p}(t)$, and

$$H_{j|i} = \begin{cases} 1 \text{ if } \Phi((< K_i)_i, X(t)) < K \text{ or} \\ \quad \Phi((< K_i)_i, X(t)) \geq K \text{ and } j \in \{j | \Phi((< K_i)_i, (< K_j)_j, X(t)) \geq K\}. \\ 0 \text{ otherwise} \end{cases}$$

Equation (3.36) represents the impact of component j on system reliability, and when a component fails and needs to be repaired, components with higher priority can be selected for PM, thereby maximising the reliability of the system.

The CMP can be used to prioritise components in binary systems while a failed component is being repaired. However, for multi-state systems, determining the priority or state of components is more complicated. Because the performance of a multi-state system can be measured by using the performance utility or simply by using the degradation status of states. In the following, we consider both cases of multistate systems.

Case I. Immediately after the system state degrades to a state below K, the system needs maintaining.

Case II. Maintenance of the system is only required if the system has degraded K states (where $k > 1$). We assume that maintenance is imperfect, i.e. the system cannot be restored to a perfect state.

3.2.1 *Priority Under Case I*

Suppose that state K_i is the threshold state of component i. That is, once the state of a component degrades to a state below K_i, a certain symptom of performance immediately appears and can be detected, namely, the degradation from one state to another is self-announcing. Assume the detected state after maintenance is $(K_0)_i$, when the state of component i is below K_i.

Definition 3.4 If component i has degraded to a state worse than K_i, the CMP of component j is defined by

$$I_{j|i}^{M}(t) = H_{j|i} I_{j|i}(t), \tag{3.37}$$

and

$$\begin{aligned} I_{j|i}(t) = \sum_{v=1}^{M} (a_v - a_{v-1}) [&\Pr\left(\Phi\left((K_o)_i, X_j(t), \boldsymbol{X}(t)\right) \geq v\right) \\ &- \Pr\left(\Phi\left((K_o)_i, X_j(t) - 1, \boldsymbol{X}(t)\right) \geq v\right)], \end{aligned} \tag{3.38}$$

where $(< K_i)_i$ represents that the state of component i degrades to a state below its threshold state K_i. $\Phi\left((< K_i)_i, \boldsymbol{X}(t)\right) < K$ represents the system fails, and component i is a critical component. $H_{j|i}$ ensures that critical components will not be selected for PM, given that component i is non-critical.

Denote

$$\mathbf{X}_i = \left((k_1)_1, (k_2)_2, \ldots, (k_{i-1})_{i-1}, *, (k_{i+1})_{i+1}, \ldots, (k_n)_n\right),$$

and

$$\mathbf{X}_{ij} = \left((k_1)_1, \ldots, (k_{i-1})_{i-1}, *, (k_{i+1})_{i+1}, \ldots \left(k_{j-1}\right)_{j-1}, *, \left(k_{j+1}\right)_{j+1}, \ldots, (k_n)_n\right). \tag{3.39}$$

$I_{j|i}(t)$ is the importance of component j given that component i's state has downgraded. Below we give an example to show how the CMP works.

Example 3.1 Suppose a multi-state system is composed of four multi-state components with the following system structure function

$$\Phi(\boldsymbol{X}(t)) = \Phi\left(X_1(t), X_2(t), X_3(t), X_4(t)\right)$$
$$= \min\{\max\left\{X_1(t), X_2(t), X_3(t)\right\}, X_4(t)\}. \tag{3.40}$$

Suppose both the state space of each component and that of the system are {0, 1, 2}. Assume both the performance values of each component and the system are 1, which means that when the state of component/system are smaller than 1, the corresponding component/system will fail. Then we have the following results.

- The case that component 4 degrades to a state below 1.
  If component 4 has degraded to a state below 1, according to the structure function, we have $\Phi(\boldsymbol{X}(t)) < 1$. Then $H_{j|4} = 1$ and

$$I_{j|4}^{M}(t) = I_{j|4}(t)$$
$$= \sum_{v=1}^{M} a_v \left[\Pr\left(\Phi\left((K_o)_4, X_j(t), \boldsymbol{X}(t)\right) = v\right)\right.$$
$$\left.- \Pr\left(\Phi\left((K_o)_4, X_j(t) - 1, \boldsymbol{X}(t)\right) = v\right)\right]. \tag{3.41}$$

  As an example, we show how to computer $I_{1|4}^{M}(t)$ below.

  Assume $(K_o)_4 = 2,\ X_1(t) = 1$, then

$$I_{1|4}^{M}(t) = \sum_{v=1}^{2} a_v \left[\Pr\left(\Phi\left((2)_4, 1_1, \boldsymbol{X}(t)\right) = v\right) - \Pr\left(\Phi\left((2)_4, 0_1, \boldsymbol{X}(t)\right) = v\right)\right]. \tag{3.42}$$

  Let

$$p_{im} = \Pr\left\{X_i(t) = m\right\},$$

then

$$\Pr\left(\Phi\left((2)_4, 1_1, \boldsymbol{X}(t)\right) = 1\right) = p_{21}(t)p_{31}(t) + p_{21}(t)p_{30}(t)$$
$$+ p_{20}(t)p_{31}(t) + p_{20}(t)p_{30}(t),$$

and

$$\Pr\left(\Phi\left((2)_4, 0_1, \boldsymbol{X}(t)\right) = 1\right) = p_{21}(t)p_{31}(t) + p_{21}(t)p_{30}(t) + p_{20}(t)p_{31}(t).$$

Thus, we have

$$a_1 \left[\Pr\left(\Phi\left((2)_4, 1_1, \boldsymbol{X}(t)\right) = 1\right) - \Pr\left(\Phi\left((2)_4, 0_1, \boldsymbol{X}(t)\right) = 1\right)\right] = a_1 p_{20}(t) p_{30}(t). \tag{3.43}$$

Besides,

$$\Pr\left(\Phi\left((2)_4, 1_1, \boldsymbol{X}(t)\right) = 2\right) = \Pr\left(\Phi\left((2)_4, 0_1, \boldsymbol{X}(t)\right) = 2\right) = p_{22}(t) + p_{32}(t), \tag{3.44}$$

so we have

$$a_2 \left[\Pr\left(\Phi\left((2)_4, 1_1, \boldsymbol{X}(t)\right) = 2\right) - \Pr\left(\Phi\left((2)_4, 0_1, \boldsymbol{X}(t)\right) = 2\right)\right] = 0. \tag{3.45}$$

We can then obtain

$$I_{1|4}^M(t) = a_1 p_{20}(t) p_{30}(t). \tag{3.46}$$

- The case that component 1 degrades to a state below 1 and the states of the other components are higher than 1.
  If component 1 is the only component that has degraded to a state below 1, according to the structure function, we have $\Phi(\boldsymbol{X}(t)) \geq 1$. Then $I_{4|1}^M(t) = 0$. Thus, one of components 2 and 3 can be selected for PM. We take component 2 for example to show how to computer $I_{2|1}^M(t)$ below.

Assume $(K_o)_1 = 0$ and $X_2(t) = 1$, then

$$I_{2|1}^M(t) = \sum_{v=1}^{2} a_v \left[\Pr\left(\Phi\left((0)_1, 1_2, \boldsymbol{X}(t)\right) = v\right) - \Pr\left(\Phi\left((0)_1, 0_2, \boldsymbol{X}(t)\right) = v\right)\right]. \tag{3.47}$$

We have:

$$\begin{aligned} \Pr\left(\Phi\left((0)_1, 1_2, \boldsymbol{X}(t)\right) = 1\right) = & p_{41}(t) p_{30}(t) + p_{41}(t) p_{31}(t) + p_{41}(t) p_{32}(t) \\ & + p_{42}(t) p_{30}(t) + p_{42}(t) p_{31}(t), \end{aligned} \tag{3.48}$$

and

$$\Pr\left(\Phi\left((0)_1, 0_2, \boldsymbol{X}(t)\right) = 1\right) = p_{41}(t) p_{31}(t) + p_{41}(t) p_{32}(t) + p_{42}(t) p_{31}(t). \tag{3.49}$$

Hence,

$$\begin{aligned} & a_1 \left[\Pr\left(\Phi\left((0)_1, 1_2, \boldsymbol{X}(t)\right) = 1\right) - \Pr\left(\Phi\left((0)_1, 0_2, \boldsymbol{X}(t)\right) = 1\right)\right] \\ = & a_1 \left[p_{41}(t) p_{30}(t) + p_{42}(t) p_{30}(t)\right]. \end{aligned} \tag{3.50}$$

Since $\Pr(\Phi((0)_1, 1_2, \boldsymbol{X}(t))=2) = \Pr(\Phi((0)_1, 0_2, \boldsymbol{X}(t)) = 2) = p_{42}(t)p_{32}(t)$. We have $a_2\left[\Pr(\Phi((0)_1, 1_2, \boldsymbol{X}(t)) = 2) - \Pr(\Phi((0)_1, 0_2, \boldsymbol{X}(t)) = 2)\right] = 0$, and $I^M_{2|1}(t) = a_1[p_{41}(t)p_{30}(t) + p_{42}(t)p_{30}(t)]$.

In the following, we investigate two scenarios and give the corresponding expressions of $I_{j|i}(t)$ to analyse the effect of component j on the system performance while component i is being maintained in a multistate system. Denote the threshold state of component j $(j \neq i)$ by K_j.

Scenario 1. The state of the component that leads to system state degrading to a state below K can be observed, but other component states cannot be detected. Assume the state degradation of component i leads to the system state worse than K. Let the observed state of component i be $(K_o)_i$. Similarly to Eq. (3.34), we have

$$\begin{aligned} &U\left((K_o)_i, \boldsymbol{X}(t)\right) \\ &= \sum_{v=1}^{M} a_v \Pr\left(\Phi\left(X_1(t), \ldots, X_{i-1}(t), K_o, X_{i+1}(t), \ldots X_n(t)\right) = v\right). \end{aligned} \tag{3.51}$$

Based on Eq. (3.2), the CMP of component j is

$$\begin{aligned} I_{j|i}(t) &= \frac{\partial U\left((K_o)_i, \boldsymbol{X}(t)\right)}{\partial \rho_{jK_j}(t)} \\ &= \sum_{v=1}^{M} (a_v - a_{v-1}) \left[\Pr\left(\Phi\left((K_o)_i, K_j, \boldsymbol{X}(t)\right) \geq v\right)\right. \\ &\quad \left. - \Pr\left(\Phi\left((K_o)_i, (K-1)_j, \boldsymbol{X}(t)\right) \geq v\right)\right]. \end{aligned} \tag{3.52}$$

Equation (3.52) describes the effect of component j on the system performance when component i is being maintained under Scenario 1.

Scenario 2. Assume component i degrades, which causes the system degrade to a state worse than K. The state of a component in the system can be detected. Let the observed state of component i be $(K_o)_i$. Similarly to Eq. (3.35), we can use Eq. (3.53) to analyse the effect of component j on the system performance while component i is being maintained. The CMP of component j is

$$\begin{aligned} I_{j|i}(t) &= \frac{\partial U\left((K_o)_i, \boldsymbol{X}(t)\right)}{\partial \rho_{j(K_o)_j}(t)} \\ &= \sum_{v=1}^{M} (a_v - a_{v-1}) \left[\Pr\left(\Phi\left((K_o)_i, (K_o)_j, \boldsymbol{X}(t)\right) \geq v\right)\right. \\ &\quad \left. - \Pr\left(\Phi\left((K_o)_i, (K_o-1)_j, \boldsymbol{X}(t)\right) \geq v\right)\right]. \end{aligned} \tag{3.53}$$

3.2.2 Priority Under Case II

Based on the Case II, we have the following scenarios.

Scenario 3. Both the system state and component states can be detected at the same time.

Suppose the system fails when the state degradation of component i causes the system to jump to the kth state. We define the following measure that prioritises the components to be maintained during the time of a failed component being repaired.

Definition 3.5 If component i has failed, the CMP of component j is defined by

$$I_{j|i}^{M}(t) = H_{j|i} I_{j|i}(t). \tag{3.54}$$

$D_{X_{i'}(t) \rightarrow X_i(t)}$ represents that the state of component i degrades from state $X_{i'}(t)$ to $X_i(t)$, and $D_{X_{j'}(t) \rightarrow X_j(t)}$ represents that the state of component j degrades from state $X_{j'}(t)$ to $X_j(t)$. Suppose component i's degrading from state $X_{i'}(t)$ to $X_i(t)$ causes the value of the function $\phi(.)$ to reduce for more than k states, then a maintenance is triggered, otherwise, no action will be taken. $I_{j|i}(t)$ is the importance of component j, given that component i's state has downgraded.

Scenario 4. The degradation of a system can be detected, but the state of a component cannot be detected.

Since the state degradation of a component cannot be detected, we cannot determine which component caused the system to jump to the k states. Here we use the effect of all states of a component on the system performance.

To measure the expected absolute deviation of the system reliability caused by a particular component's different performance levels and associated probabilities. Ramirez-Marquez and Coit [3] gave the following alternative composite importance measure, or mean absolute deviation (MAD),

$$I_i^{mad}(t) = \sum_m p_{im}(t) \left| \Pr\left(\Phi\left(m_i, \boldsymbol{X}(t)\right) \geq d\right) - \Pr\left(\Phi\left(\boldsymbol{X}(t) \geq d\right)\right) \right|, \tag{3.55}$$

where d is a constant system requirement, and $p_{im}(t)$ is the probability that component i is in state m at time t. $I_i^{mad}(t)$ is the expected absolute deviation of component i from the reliability of the system.

Based on the expected performance of a system, $U(\boldsymbol{X}(t))$, and pre-specified performance utility threshold (i.e., w), we can obtain the expected absolute deviation of component i in the system performance.

$$I_i^{U}(t) = \sum_m p_{im}(t) \left| U\left(m_i, \mathbf{X}_i(t)\right) \geq w - U\left(\boldsymbol{X}(t)\right) \geq w \right|. \tag{3.56}$$

$I_i^U(t)$ represents the expected absolute deviation of component i for system performance based on the work by Ramirez-Marquez and Coit [3]. Let $I_{i^*}^U(t) = \max\{I_i^U(t)\}$, where the corresponding component of $I_{i^*}^U(t)$ is denoted as i^*. As such, we introduce the following definition.

Definition 3.6 If component i^* has failed, the CMP of component j is defined by

$$I_{j|i^*}^M(t) = H_{j|i^*} \frac{I_{i^*}^U(t)}{\sum_i I_i^U(t)} I_{j|i^*}(t) \tag{3.57}$$

where $\frac{I_{i^*}^U(t)}{\sum_i I_i^U(t)}$ represents the ratio of component i^* in the expected absolute deviations of all components.

3.2.3 Linking Maintenance Policies

This section analyses how to identify the components of PM in a maintenance policy.

3.2.3.1 Maintenance Policies Under Case I

Maintenance policy A. Once a component degrades below its threshold state, the component must be maintained. Under this strategy, the components being maintained may be critical or non-critical. There are two cases:

- If a critical component fails, the system fails. PM can be conducted on other components.
- If a non-critical component fails, the system can continue working. PM can be performed on other non-critical components.

If the state of component i has degraded below its threshold state K_i, then under maintenance policy A, the CMP of component j is defined as

$$I_{j|i}^M(t) = H_{j|i} I_{j|i}(t). \tag{3.58}$$

The symbol $(< K)_i$ represents that the state of component i degrades to a state below its threshold state K_i. If component i degrades to a state below K_i and causes the system to reduce to a state below its threshold state K, i.e. $\Phi\left((< K)_i, \boldsymbol{X}(t)\right) < K$, then component i is critical and the system stops working. Thus, the PM can be performed on all other components, $j \in \{1, \ldots, i-1, i+1, \ldots, n\}$. If $\Phi\left((< K)_i, \boldsymbol{X}(t)\right) \geq K$, then component i is non-critical. Thus, PM can be performed on the non-critical components, $j \in \left\{j | \Phi\left((< K)_i, (< K)_j, \boldsymbol{X}(t)\right) \geq K\right\}$. For maintenance policy A, Scenarios 1 and 2 are suitable for replacing Eqs. (3.53) and (3.54) with Eq. (3.59). When repairing component i, component j with the largest $I^M_{j|i}(t)$ should be selected first for PM to improve system performance. Then, the engineer may select components for PM in light with the order of importance measures $I^M_{j|i}(t)$.

Maintenance policy B. A system fails when the degradation of some components causes the system state to degrade below the threshold K. Maintenance is therefore needed. The corresponding components can be detected. The components that are maintained under this policy include critical and non-critical components. Assume that the set of degraded components that cause the system state to degrade below the threshold state K is $\{i_1, i_2, \ldots, i_m\}$. The set of components $\{i_1, i_2, \ldots, i_m\}$ is the cut set of the system. Under maintenance policy B, when components $\{i_1, i_2, \ldots, i_m\}$ are being maintained, the system stops working and PM can be executed on all other components.

Under Scenario 1, when the state of component $\{i_1, i_2, \ldots, i_m\}$ is $(K_o)_{i_1}$, $(K_o)_{i_2}, \ldots, (K_o)_{i_m}$, and there is no other component state corresponding to the performance observed, the CMP of component j is

$$\begin{aligned} I^M_{j|i_1,i_2,\ldots,i_m}(t) &= I_{j|i_1,i_2,\ldots,i_m}(t) \\ &= \frac{\partial U\left((K_o)_{i_1}, (K_o)_{i_2}, \ldots, (K_o)_{i_m}, \boldsymbol{X}(t)\right)}{\partial \rho_{jK_j}(t)} \\ &= \sum_{v=1}^{M} (a_v - a_{v-1}) \left[\Pr\left(\Phi\left((K_o)_{i_1}, (K_o)_{i_2}, \ldots, (K_o)_{i_m}, K_j, \boldsymbol{X}(t)\right) \geq v\right)\right. \\ &\quad \left. - \Pr\left(\Phi\left((K_o)_{i_1}, (K_o)_{i_2}, \ldots, (K_o)_{i_m}, (K-1)_j, \boldsymbol{X}(t)\right) \geq v\right)\right]. \end{aligned} \tag{3.59}$$

In Scenario 2, when the state of component $i_1, i_2, \ldots, i_m$ is $(K_o)_{i_1}, (K_o)_{i_2}, \ldots, (K_o)_{i_m}$, if the performance corresponding to the state of other components can be observed, then the CMP of component j is

$$\begin{aligned} I^M_{j|i_1,i_2,\ldots,i_m}(t) &= I_{j|i_1,i_2,\ldots,i_m}(t) \\ &= \frac{\partial U\left((K_o)_{i_1}, (K_o)_{i_2}, \ldots, (K_o)_{i_m}, \boldsymbol{X}(t)\right)}{\partial \rho_{j(K_o)_j}(t)} \\ &= \sum_{v=1}^{M} (a_v - a_{v-1}) \left[\Pr\left(\Phi\left((K_o)_{i_1}, (K_o)_{i_2}, \ldots, (K_o)_{i_m}, (K_o)_j, \boldsymbol{X}(t)\right) \geq v\right)\right. \\ &\quad \left. - \Pr\left(\Phi\left((K_o)_{i_1}, (K_o)_{i_2}, \ldots, (K_o)_{i_m}, (K_o-1)_j, \boldsymbol{X}(t)\right) \geq v\right)\right]. \end{aligned} \tag{3.60}$$

Components $\{i_1, i_2, \ldots, i_m\}$ in the system degrades to a below-threshold state and the system stops working. When maintenance is performed on component $i_1, i_2, \ldots, i_m$, the component with the maximum $I^M_{j|i_1,i_2,\ldots,i_m}(t)$ should be selected for PM so that the performance of the system can be improved to the maximum.

3.2.3.2 Maintenance Policies Under Case II

Maintenance policy C. Assume that the system fails if it has degraded k states. Thus the system need repairing. For different components, repairmen may have different capacity. When component i is being maintained, r_i states of component i can be increased. Assume one component state degradation leads to the system state degrading k states, then components for PM can be determined by $H_{j|i}$.

Based on Eq. (3.2)

$$I_{i_m}^G(t) = \sum_{v=1}^{M} (a_v - a_{v-1}) \left[\Pr\left(\Phi\left(m_i, \boldsymbol{X}(t)\right) \geq v\right) - \Pr\left(\Phi\left((m-1)_i, \boldsymbol{X}(t)\right) \geq v\right)\right]. \tag{3.61}$$

$I_{i_m}^G(t)$ represents the change in system performance as the component i changes from state $m-1$ to state m. Then the change in system performance as component i improves from state m to state $m + r_i$ is

$$\begin{aligned}
& I_{i_{(m \to m+r_i)}}^G(t) \\
&= I_{i_{(m+r_i)}}^G(t) + I_{i_{(m+r_i-1)}}^G(t) + \ldots + I_{i_{(m+1)}}^G(t) \\
&= \sum_{q=m+1}^{m+r_i} \sum_{v=1}^{M} (a_v - a_{v-1}) \left[\Pr\left(\Phi\left(q_i, \boldsymbol{X}(t)\right) \geq v\right) - \Pr\left(\Phi\left((q-1)_i, \boldsymbol{X}(t)\right) \geq v\right)\right] \\
&= \sum_{v=1}^{M} (a_v - a_{v-1}) \sum_{q=m+1}^{m+r_i} \left[\Pr\left(\Phi\left(q_i, \boldsymbol{X}(t)\right) \geq v\right) - \Pr\left(\Phi\left((q-1)_i, \boldsymbol{X}(t)\right) \geq v\right)\right] \\
&= \sum_{v=1}^{M} (a_v - a_{v-1}) \left[\Pr\left(\Phi\left(m_i + r_i, \boldsymbol{X}(t)\right) \geq v\right) - \Pr\left(\Phi\left(m_i, \boldsymbol{X}(t)\right) \geq v\right)\right] \\
&= \sum_{v=1}^{M} a_v \left[\Pr\left(\Phi\left(m_i + r_i, \boldsymbol{X}(t)\right) = v\right) - \Pr\left(\Phi\left(m_i, \boldsymbol{X}(t)\right) = v\right)\right].
\end{aligned} \tag{3.62}$$

When component i is being maintained, r_i states can be restored on it. So under Scenario 3, when component i has failed, we assume the observed state of component j is $(K_o)_j$. Then we have

$$\begin{aligned}
I_{j|i}(t) = & \sum_{v=1}^{M} a_v \Pr\left(\Phi\left((K_o)_i, (K_o)_j + r_j, \boldsymbol{X}(t)\right) = v\right) \\
& - \sum_{v=1}^{M} a_v \Pr\left(\Phi\left((K_o)_i, (K_o)_j, \boldsymbol{X}(t)\right) = v\right).
\end{aligned} \tag{3.63}$$

If component i has failed, the CMP of component j is

$$I_{j|i}^{M}(t) = H_{j|i} \sum_{v=1}^{M} a_v \Pr\left(\Phi\left((K_o)_i, (K_o)_j + r_j, \boldsymbol{X}(t)\right) = v\right) - H_{j|i} \sum_{v=1}^{M} a_v \Pr\left(\Phi\left((K_o)_i, (K_o)_j, \boldsymbol{X}(t)\right) = v\right). \tag{3.64}$$

In Scenario 4, component states cannot be observed. Then we have

$$I_{j|i^*}(t) = \sum_{v=1}^{M} a_v \left[\Pr\left(\Phi\left((K)_{i^*}, K_j + r_j, \boldsymbol{X}(t)\right) = v\right) - \Pr\left(\Phi\left((K)_{i^*}, K_j, \boldsymbol{X}(t)\right) = v\right)\right]. \tag{3.65}$$

If component i^* has failed, the CMP of component j is defined by

$$I_{j|i^*}^{M}(t) = H_{j|i^*} \frac{I_{i^*}^{U}(t)}{\sum_i I_i^{U}(t)} \sum_{v=1}^{M} a_v \left[\Pr\left(\Phi\left((K)_{i^*}, K_j + r_j, \boldsymbol{X}(t)\right) = v\right) - \Pr\left(\Phi\left((K)_{i^*}, K_j, \boldsymbol{X}(t)\right) = v\right)\right]. \tag{3.66}$$

3.2.4 When Maintenance Budget Is Limited

Based on a fixed maintenance budget C, we can determine the composition of the PM in order to maximise the expected system performance at time t.

(1) When the maintenance cost of each component is the same, components for PM can be determined by $I_{j|i}^{M}(t)$ and $I_{j|i_1,i_2,\ldots,i_m}^{M}(t)$ based on the component importance ranking.
(2) In the case where the PM costs of different components are different, components with greater importance incur greater PM costs. In this case, prioritising PM allocation to components with larger importance measures based on $I_{j|i}^{M}(t)$ and $I_{j|i_1,i_2,\ldots,i_m}^{M}(t)$ is not always the optimal allocation method. Therefore, we should use the integer programming models in the following to determine the components for PM.

Then under Maintenance Policies A and C, when component i undergoes repair, we use a fixed time t to solve the following equation.

$$\max_{z_j} \sum_{j \neq i} z_j I_{j|i}^{M}(t), \tag{3.67}$$

subject to

$$c_i + \sum_{j \neq i} c_j z_j \leq C$$

and

$$z_j \in \{0, 1\},$$

in which c_i is the repair cost for component i, c_j represents the maintenance cost for component j and z_j is the decision variable representing whether component j should be maintained or not. Note that z_j can only take values from 0 and 1.

When components $i_1, i_2, \ldots, i_m$ are being repaired, in order to resolve the conflict between the expected cost ratio in PM and importance-based PM maintenance order, we use an integer programming method within the fixed time t to solve the following problem:

$$\max_{z_j} \sum_{j \neq i_1, i_2, \ldots, i_m} I^M_{j|i_1, i_2, \ldots, i_m}(t) \cdot z_j, \tag{3.68}$$

subject to

$$c_{i_1} + c_{i_2} + \ldots + c_{i_m} + \sum_{j \neq i_1, i_2, \ldots, i_m} c_j z_j \leq C \tag{3.69}$$

and

$$z_j \in \{0, 1\}, \tag{3.70}$$

where $c_{i_1}, c_{i_2}, \ldots, c_{i_m}$ are the repair costs of components $i_1, i_2, \ldots, i_m$, respectively.

For the above integer programming model, we assume that the optimal maintenance policy is $\left\{z_j^*, j \neq i\right\}$ and $\left\{z_j^*, j \neq i_1, i_2, \ldots, i_m\right\}$, then the set of optimal components for PM is $\left\{j | z_j^* = 1\right\}$. Actually, $\sum_{j \neq i} z_j^*$ and $\sum_{j \neq i_1, i_2, \ldots, i_m} z_j^*$ are the number of maintained components.

In addition, if maintenance time is considered, then it is assumed that the PM time on a working component is less than the repair time of a failed component. Otherwise, PM will delay the system operation and thus reduce the system performance.

3.2.5 Preventive Maintenance Strategies Considering Environmental Importance

Suppose that the components of a system do not fail catastrophically, but degrade over time. That is, the critical state of a component usually changes over time and is affected by environmental conditions. However, existing studies usually ignore the effect of uncertainty in the external environment on system reliability. With this in mind, this section identifies the relationship between the degradation process of a component and the environmental conditions, and proposes an environmental importance measure. Based on this, a PM strategy based on the environmental importance measure is proposed.

3.2.5.1 Degradation modelling in External Shocks

The system under investigation is assumed to be composed of n components, which are statistically independent of each other. When any degradation process k_i relates to component i reaches the threshold level $n_j^{(i)}$, component i fails. That is, component i has multiple competing failure modes due to multi-dimensional degradation. The degradation process of components is modelled with a K-dimensional Wiener process. The degradation process j of component i is denoted by $\left(\mathbf{X}_j^{(i)}\right)_{j=1,2,\ldots,k}$. Then the relationship between the degradation process and external shocks is established, and the expression of environmental importance is obtained.

Scenario 1. Deterministic environmental condition Let $e_t : [0, \infty) \rightarrow R$ be a time-dependent real-valued function that specifies the external shock at time t. Under certain external shocks, $e_t = e_0, t \geq 0$, the degradation process j of component i is modelled by:

$$d\mathbf{X}_j^{(i)}(t; e_0) = \mu_{j,0}^{(i)} dt + \sigma_{j,0}^{(i)} dB_t. \tag{3.71}$$

where $(B_t)_{t\geq 0}$ is the standard Brownian motion, and $\mu_{j,0}^{(i)}$ and $\sigma_{j,0}^{(i)}$ are the degradation rate and diffusion coefficient under a constant external shock, e_0, respectively. In practice, the degradation rate under a constant environmental condition is often determined by the physics-of-failure and can be modelled by degradation testing. The diffusion coefficient refers to the expected value of degradation change per unit of time. The generalized Wiener process has a constant expected diffusion coefficient.

The function $k_j^{(i)}(e_t)$ models the influence of the external shocks on both the degradation rate and diffusion coefficient:

$$k_j^{(i)}(e_t) = \frac{\mu_j^{(i)}(t, e_t)}{\mu_{j,0}^{(i)}} = \left(\frac{\sigma_j^{(i)}(t, e_t)}{\sigma_{j,0}^{(i)}}\right)^2, \tag{3.72}$$

with $\mu_j^{(i)}(t, e_t)$ and $\sigma_j^{(i)}(t, e_t)$, respectively, being the degradation rate and diffusion coefficient of the degradation process j of component i at time t and under the external shock e_t. The choice of $k_j^{(i)}(e_t)$ is always case dependent in practice. For example, the degradation of electronic devices can often be attributed to the free energy difference between the initial state and the degraded state, and the function $k_j^{(i)}(e_t)$ becomes the Arrhenius equation. However, in the Arrhenius model, only the influence of a single temperature stress on the change of physical and chemical properties of the system is considered. In engineering practice, multiple stresses are acting on the system at the same time. While the Eyring model belongs to the multi-stress model, it is more general to use the Eyring model to represent external shocks $k_j^{(i)}(e_t)$:

$$k_j^{(i)}(e_t) = K(T, S) = \frac{dM}{dt} = A\frac{kT}{h}e^{-\frac{k}{ET}}e^{S(C+\frac{D}{kT})} = K_0 f_1 f_2, \tag{3.73}$$

where T is the temperature stress (thermodynamic temperature), S is the non-temperature stress, $\frac{dM}{dt}$ is the chemical reaction rate, $K_0 = A\frac{kT}{h}e^{-k/ET}$ is the Eyring reaction rate with only temperature stress, h is the Planck constant, E is the activation energy (obeying Boltzmann distribution), k is the Boltzmann constant, $f_1 = e^{SC}$ is the correction factor for the energy distribution in the presence of non-temperature stresses, and $f_2 = e^{DS/kT}$ is the correction factor for the activation energy in the presence of non-temperature stresses.

Let $x_j^{(i)}(0)$ be the initial degradation level of the degradation process j of component i. Thus, the degradation process under a time-varying external shock e_t is:

$$\begin{aligned}\mathbf{X}_j^{(i)}(t; e_0) &= x_j^{(i)}(0) + \int_0^t d\mathbf{X}_j^{(i)}(x; e_0)\\ &= x_j^{(i)}(0) + \int_0^t \mu_{j,0}^{(i)}k_j^{(i)}(e_x)dx + \int_0^t \sigma_{j,0}^{(i)}(k_j^{(i)}(e_x))^{1/2}dB_x.\end{aligned} \tag{3.74}$$

According to Liu et al. [2], this process is a Wiener process with a time-dependent mean value function and diffusion.

The time when $\mathbf{X}_j^{(i)}(t; e_0)$ first attains the threshold $n_j^{(i)}$—i.e., the first passage time—is given by $T_j^{(i)} = inf(t; \mathbf{X}_j^{(i)}(t; e_0) \geq n_j^{(i)})$. In a special case when the occurrence of external shocks is independent of time, it is well known that $T_j^{(i)}$ follows an inverse Gaussian distribution. Because component i is subject to k_i competing failure modes due to multi-dimensional degradation, the lifetime of component i is defined as $T^{(i)} = \min_{j=1,2,\ldots,k_i} T_j^{(i)}$.

If e_t is deterministic, the degradation rates and diffusion coefficients of the k_i degradation processes associated with component i are also deterministic and are statistically independent. Hence, the reliability of the component i is given by:

$$R^{(i)}(t; e_t) = \Pr\left(T^{(i)} > t; e_t\right) = \prod_{i=1}^{k_i}(1 - F_{T_j^{(i)}}(t; e_t)). \tag{3.75}$$

Let $\mathbf{Z}^{(i)}(t) = 1$ when component i functions at time t, and $\mathbf{Z}^{(i)}(t) = 0$ when component i is in a failed state at time t, and $\mathbf{Z}(t) = \left(\mathbf{Z}^{(1)}(t), \mathbf{Z}^{(2)}(t), \ldots, \mathbf{Z}^{(n)}(t)\right)$. Then the system structure function, $\varphi(\mathbf{Z}(t))$, is defined as:

$$\varphi(\mathbf{Z}(t)) = \begin{cases}1, \textit{if system functions at time } t\\ 0, \textit{if system fails at time } t\end{cases}. \tag{3.76}$$

Thus, the environmental importance of component i with a multidimensional degradation process under deterministic environmental conditions can be obtained by:

$$\begin{aligned} I_i^B(t; e_t) &= \Pr\left(\varphi(\mathbf{Z}(t)) = 1 | \mathbf{Z}^{(i)} = 1; e_t\right) \\ &\quad - \Pr\left(\varphi(\mathbf{Z}(t)) = 1 | \mathbf{Z}^{(i)} = 0; e_t\right) \\ &= \frac{\partial R(t; e_t)}{\partial R^{(i)}(t; e_t)}. \end{aligned} \tag{3.77}$$

Scenario 2. Stochastic environmental condition When e_t is stochastic, the degradation rates and diffusion coefficients of the k_i degradation processes associated with component i are also stochastic. Additionally, as the degradation rates and diffusion coefficients of the k_i degradation processes share the same environmental condition e_t, the k_i degradation processes are no longer statistically independent.

Suppose that the external shock can be written as $e_t = \widehat{e_t} + b_t$, where b_t is a stochastic process and $\widehat{e_t}$ is the predicted external shock at time t. b_t is always case-dependent. For example, if b_t is a Brownian motion with drift zero and diffusion σ_e—i.e., $b_t = (\sigma_e B_t)_{t \geq 0}$ and $\varepsilon_{tn} = 0$—then e_t is a Brownian motion with a mean value function $\widehat{e_t}$ and a fluctuation b_t, or b_t can be a normal random representing the white noise.

The environmental importance of component i under random external shocks can be calculated using the following Monte Carlo method.

$$I_i^B(t; e_t) \approx \widehat{I_i^B}(t; e_t) = \frac{1}{N} \sum_{p=1}^{N} I_i^{BM}\left(t; \widetilde{e_t}^{(p)}\right). \tag{3.78}$$

3.2.5.2 Preventive Maintenance (PM) Based on Environmental Importance

Assuming that the states of the corresponding components can be observed, in order to propose PM strategies to achieve more accurate PM and maximize the expected performance of the system, this section builds on the analysis of the environmental significance presented in Sect. 3.2.5.1. Combined with the joint importance concept proposed by Dui et al. [1], the impact of component j on system reliability when component i is repaired is expressed as

$$I_{i|j}^B(t; e_t) = \frac{\partial^2 R(t; e_t)}{\partial R^{(i)}(t; e_t)\, \partial R^{(j)}(t; e_t)}. \tag{3.79}$$

Let $\mathbf{Y}_i(t)$ represent the state of component i at time t, $\mathbf{Y}_i(t) = 0, 1, 2, \ldots, M_i$, $\mathbf{Y}(t) = (\mathbf{Y}_1(t), \mathbf{Y}_2(t), \ldots, \mathbf{Y}_n(t))$ represents the state vector of the component, and $\Phi(\mathbf{Y}(t))$ is the system structure function.

Once the state of a component degrades below its threshold state, the corresponding component may need to be located and must be repaired. In this case, the components being repaired may be critical or non-critical. Assuming that the component under maintenance is critical, the system must stop working. Preventive maintenance (PM) can be performed on all other components. If the maintenance component is not critical, the system does not need to stop working. CM can be performed on non-critical components.

Assuming that the state of component i is less than its threshold state n^i, then under the above maintenance policy, let the component maintenance priority (CMP) of component $j (j \neq i)$ be:

$$I_i^M(t; e_t) = H_{j|i} \frac{\partial^2 R(t; e_t)}{\partial R^{(i)}(t; e_t) \partial R^{(j)}(t; e_t)}. \tag{3.80}$$

$$H_{j|i} = \begin{cases} 1 \text{ if } \Phi((< n)^i, \boldsymbol{Y}(t)) < N \text{ or} \\ \quad \Phi((< n)^i, \boldsymbol{Y}(t)) \geq N \text{ and } j \in \{j | \Phi((< n)^i, (< n)^j, \boldsymbol{X}(t)) \geq N\}, \\ 0 \text{ otherwise} \end{cases}$$

where $(< n)^i$ indicates that the state of component i degrades below its threshold state n^i. If the state degradation of component i falls below n^i, causing the value of the system structure function $\Phi(\bullet)$ to fall below its threshold state N, i.e. $\Phi\left((< n)^i, \mathbf{Y}(t)\right) < N$, component i becomes critical and the system stops functioning. Therefore, PM are performed on all other components $j \in \{1, \ldots, i-1, i+1, \ldots, n\}$. Component i is non-critical if $\Phi\left((< n)^i, \mathbf{Y}(t)\right) \geq N$. So PM are performed on non-critical components $j \in \left\{j | \Phi\left((< n)^i, (< n)^j, \mathbf{X}(t)\right) \geq N\right\}$.

When component i is repaired, which has the largest $I_i^M(t; e_t)$, component j should be chosen for PM.

Given limited maintenance costs, the set of components for PM should be determined to maximize the reliability of the system, given a fixed total maintenance cost of C.

$$\text{Maximize: } \sum_{j \neq i} z_j I_i^M(t; e_t) \tag{3.81}$$

$$\text{Subject to: } c^i + \sum_{j \neq i} c^j z_j \leq C \tag{3.82}$$

$$z_j \in \{0, 1\}, \tag{3.83}$$

where c^i is the maintenance cost of component i, z_j is the PM variable of component j and represents the decision variable of whether to maintain component j. z_j can only adopt the value 0 or 1. When $z_j = 1$, PM is carried out for the component j; otherwise, no maintenance is carried out.

3.3 Importance-Informed Component Maintenance Priority

Let $a_0 \leq a_1 \leq \ldots \leq a_n$ represent the performance levels corresponding to the state space $\{0, 1, \ldots, n\}$ of a multistate system. If we assume that $a_0 = 0$, without loss of generality, then the expected performance of the system can be defined by:

$$\begin{aligned} U(\boldsymbol{X}(t)) &= \sum_{v=1}^{n} a_v \Pr(\Phi(\boldsymbol{X}(t)) = v) \\ &= \sum_{v=1}^{n} a_v \Pr\left(\Phi(X_1(t), X_2(t), \ldots X_n(t)) = v\right). \end{aligned} \tag{3.84}$$

But for a multistate system, if the degradation of a component causes the performance of the system to degrade, the component may need repairing. The time on maintaining this component provides an opportunity of maintaining other components to improve the system performance. This poses a challenge on which components have the top priority for PM.

For multistate systems, we represented 2 cases:

Case I. When the system state is below K, it needs maintaining.
Case II. Only if the system degrades k states (where $k > 1$), it needs maintaining. The maintenance is assumed imperfect, that is, the system cannot be restored to the perfect state.

Based on the following two cases, we also give two kinds of component maintenance priority.

If component i has degraded to a state below k_i, the CMP of component j is defined by

$$I_{j|i}^{M}(t) = H_{j|i} I_{j|i}(t) \tag{3.85}$$

Then we give an example to show how the CMP works. Based on the Case II, we give 2 scenarios and 2 definitions.

Scenario 1: Both system state and component state can be detected.

Definition 3.7 If component i has failed, the CMP of component j is defined by

$$I_{j|i}^{M}(t) = H_{j|i} I_{j|i}(t). \tag{3.86}$$

Scenario 2: The state of the system can be detected, but the states of all component cannot be observed.

Definition 3.8 If component i has failed, the CMP of component j is defined by

$$I_{j|i}^{M}(t) = H_{j|i} \frac{I_{i*}^{U}(t)}{\sum_i I_i^U(t)}. \tag{3.87}$$

3.4 Optimise the Number of Components for Preventive Maintenance

Based on Case I, we can give two policies: Maintenance Policy A and Maintenance Policy B.

Maintenance Policy A: Once the state of a component degrades to a state below its threshold state, the component must be maintained. Under this policy, the maintained component can be critical or non-critical. There are the following two situations.

- If the maintained component is critical, then the system must stop working. PM may be performed on other components.
- If the maintained component is non-critical, then the system does not have to stop working. PM can be performed on the non-critical components. If the state of component i has degraded to a state below its threshold state k_i, then under maintenance policy A, the CMP of component j is defined by

$$I_{j|i}^{M}(t) = H_{j|i} I_{j|i}(t). \tag{3.88}$$

For maintenance policy A, Scenarios 1 and 2 are suitable by integrating equations. When component i undergoes maintenance, component j with the maximum $I_{j|i}^{M}(t)$ should be first selected for PM so that the system performance can be improved. Then we found that selecting the component order for PM following the ranking of the component $I_{j|i}^{M}(t)$ is better.

Maintenance Policy B: When the degradation of some components cause the systems state to degrade to a state below its threshold state, a maintenance activity is undertaken, and the corresponding components can be located. Under this policy, the maintained components may consist of some critical components, or some non-critical components. Assume the set of degraded components that cause the systems state to degrade into a state below its threshold state K is $\{i_1, i_2, \ldots, i_m\}$. Actually, the set of components $i_1, i_2, \ldots, i_m$ is a cut set of the system. Under maintenance policy B, when components $i_1, i_2, \ldots, i_m$ are being maintained, the system stops working, and PM can be performed on all other components.

Based on the Case II, we give maintenance policy C:

Maintenance Policy C: When the system jumps k states, the system need to be repaired. For different components, repairmen may have different capacity. When component i is being maintained, it can be increased r_i states. Assume one compo-

nent state degradation leads to the system state jumping k states. This maintained component may be critical or non-critical. So PM of other components can be determined by $H_{j|i}$.
Then we considered limited maintenance cost. Given the fixed maintenance cost C, we may determine components for PM to maximise the expected system performance.

- When each component has the same maintenance cost, the components for PM can be determined following the ranking of component importance measures by $I_{j|i}^M(t)$ and $I_{j|i_1,i_2,\ldots,i_m}^M(t)$.
- Under the situation that the cost of PM on different components differs, the component with a larger importance measure may also incur a larger PM cost. In this case, it is not always optimal to allocate PM priority to the component with large importance measures by $I_{j|i}^M(t)$ and $I_{j|i_1,i_2,\ldots,i_m}^M(t)$.

Furthermore, if maintenance time is considered, then we assume that the time of PM on components is less than the repair time of failed components. Otherwise, PM will delay the system operation, which will reduce system performance.

3.5 Summary

Because of the differences and interdependencies between components, different maintenance sequences can have significantly different impacts on the system. This chapter presented a function of system components based on the loss of system performance to define the performance-based importance measures for optimisation of maintenance policies. Then, for reliability-based importance measures for optimisation of maintenance policies, the chapter presented four assumptions and component maintenance priority (CMP) to prioritise components in binary systems. It also considered both cases of multistate systems. In the end, new importance measures were given.

References

1. Dui H, Wu S, Zhao J (2021) Some extensions of the component maintenance priority. Reliab Eng Syst Saf 214:107729
2. Liu X, Al-Khalifa KN, Elsayed EA, Coit DW, Hamouda AS (2014) Criticality measures for components with multi-dimensional degradation. IIE Trans 46(10):987–998
3. Ramirez-Marquez JE, Coit DW (2007) Multi-state component criticality analysis for reliability improvement in multi-state systems. Reliab Eng Syst Saf 92(12):1608–1619
4. Wu S, Chen Y, Wu Q, Wang Z (2016) Linking component importance to optimisation of preventive maintenance policy. Reliab Eng Syst Saf 146:26–32

Chapter 4
Importance Measures for Optimisation of Cost-Based Maintenance Policies

Abstract This chapter incorporates the relevant cost information of system failures and maintenance in importance measures, for which cost-based importance measures are introduced and applied in the optimisation of maintenance policies.

Keywords Cost-based importance measure · Maintenance policy · Optimisation

When a component fails and is then being repaired, preventive maintenance (PM) may be performed on some other components in the system. The expected cost of the failure caused by the component failure relates to the reliability of the failed component. This chapter investigates component maintenance policies based on the expected system maintenance cost and the number of repairmen. First, it proposes an importance measure based on maintenance cost of different maintenance policies and discusses the integrated importance measures. A cost-based PM policy and a cost-based opportunistic PM policy are proposed to identify the component or component groups that can be selected for PM. Then three different maintenance cost scenarios are analysed for different maintenance policies. Finally, it considers the joint effects of maintenance cost and maintenance time, a cost-based integrated importance measure (IIM) to identify which component/group of components may be selected for PM. The optimisation of different PM policies by means of the use of cost-based IIM is discussed.

4.1 Cost-Based Importance Measures for Optimisation of Preventive Maintenance (PM) Policies

4.1.1 Literature Review for Maintenance

In activities such as system reliability improvement, important measures in reliability engineering can provide valuable information, so as to continuously optimise the system and can be applied to different scenarios. At the system maintenance stage,

© The Author(s), under exclusive license to Springer Nature Switzerland AG 2024
H. Dui and S. Wu, *Importance-Informed Reliability Engineering*, Springer Series in Reliability Engineering, https://doi.org/10.1007/978-3-031-52455-4_4

the use of important measures can greatly reduce maintenance costs. In order to extend the life of the system at the lowest possible cost, PM is required.

In terms of maintenance policy optimisation, an in-depth understanding of system reliability helps design more meaningful maintenance policies [3]. Ahmadi and Wu [1] proposed an integrated model for maintenance policy planning of a parallel system, assuming that the failure of the system was detected through inspection. Annamraju et al. [2] proposed a method that can better determine the time range to evaluate the reliability of a repairable system. Gao et al. [9] proposed the joint optimisation of batch and maintenance policies under two failure models. Wu et al. [23] further analysed the optimisation of maintenance policies using the concept of risk summarisation, and proposed a maintenance policy that was optimised for a group of different systems. Liu et al. [16] proposed a multi-state dynamic selection and maintenance optimisation method based on deep learning.

When a component fails, corrective maintenance needs to be performed to restore its function. In order to improve system performance or availability, it is desired to perform PM on other working components while the failed component is being repaired. However, because of the constraint budget and maintenance resources in practice, it is often not possible to perform PM on all remaining system components. In addition, different choices of components for PM can incur different costs. Therefore, it is essential to develop appropriate measures to guide the selection of components for PM in order to minimise the cost impact during the selection process.

In recent years, many papers have been published to minimise reliability-related costs. Wu and Coolen [21] proposed a new importance measure that considers component repair costs and system repair costs. Wu et al. [22] proposed an importance of component maintenance priority to determine which components should be selected for PM. Furthermore, considering the cost and time of component maintenance, Dui et al. [7] proposed a cost-based integrated importance measure to identify components or groups of components that can be selected for PM. These references focus on analysing the impact of component or component maintenance priority on binary system costs. Peng and van Houtum [17] proposed a joint optimisation model for condition-based maintenance and economic manufacturing quantity. Zhang and Zeng [29] proposed a joint optimal maintenance policy, which is accomplished by using the semi-regenerative process theory by considering the cost of maintenance and cost of managing the spare parts inventory. Yeh [27] also proposed a new method for reliability evaluation of multi-state networks under cost constraints.

Gao et al. [10] gave a conditional reliability importance to decide which components need more attention in terms of maintenance. Wu et al. [22] overcame the defects of reference [10], and proposed an importance measure of component maintenance priority to determine which components may be selected for PM so that the reliability of the system can be maximally improved, while Gao et al. [10], Wu et al. [22] ignored the maintenance time and maintenance cost.

4.1.2 A Cost-Based Component Maintenance Importance

4.1.2.1 Cost-Based Component Maintenance Importance in Binary Systems

Suppose that only the failure of a critical component can cause the system to fail, and then incurs costs $c_{s,i}$ due to system failure and maintaining component i. If the failure of a certain component does not cause the system to failure, it will only incur the cost of repairing the failed component. The total expected maintenance cost within time interval $(0, t)$ is therefore given by

$$C(t) = \sum_{i=1}^{n} \{ \{ c_{s,i} \Pr[\phi(0_i, \mathbf{1}) = 0] + c_i + C_i^P(t) \} \Pr[x_i = 0] \}, \tag{4.1}$$

where $c_{s,i}$ is the cost per system failure due to the failure to component i and c_i is the cost due to the failure of component i. $\Pr[\phi(0_i, \boldsymbol{X}(t)) = 0]$ is the probability that the system is at state 0 after component i fails. $\Pr[x_i = 0]$ is the probability that the component i fails.

The cost of PM on other components is therefore obtained by

$$\begin{aligned} C_i^P(t) = & H_{s,i} \sum_{j=1, i \neq j}^{n} c_{p_j} \Pr\left[x_j = 1\right] \\ & + (1 - H_{s,i}) \sum_{z=1}^{m} c_{p_{jz}} \Pr\left[x_{j_z} = 1\right] \Pr\left[\phi\left(0_i, 0_{j_1}, \ldots, 0_{j_{z-1}}, \mathbf{1}_{i, j_1, j_2, \ldots, j_{z-1}}\right) = 1\right], \end{aligned} \tag{4.2}$$

where $H_{s,i} = \begin{cases} 1, & \text{if } \phi(0_i, \boldsymbol{X}(t)) = 0 \\ 0, & \text{if } \phi(0_i, \boldsymbol{X}(t)) = 1, \end{cases}$ which implies that if a critical component fails, PM will be performed on all other components. Otherwise, if a non-critical component fails, the number of other components that can be maintained is limited. The maximum number of components for PM is m. $(0_i, 0_{j_1}, \ldots, 0_{j_{z-1}}, 1_{i, j_1, j_2, \ldots, j_{z-1}})$ denotes the case that components $i, j_1, j_2, \ldots, j_{z-1}$ stop working and the other components are working.

Definition 4.1 (Cost-based Component Maintenance Importance (CCMI)) If component i has failed, the CCMI of component j is defined by

$$I_{j|i}^{C} = -H_{j|i} \frac{\partial C(0_i, p(t))}{\partial p_j(t)}. \tag{4.3}$$

Below we investigate the expected total maintenance cost function based on the failure rate $\lambda_i(t)$, considering the cost function defined by Wu and Coolen [21]

$$C^{'}(t) = \sum_{j=1}^{m_0} c_{s,k_j} \wedge U_{k_j}(t) + \sum_{i=1}^{n} \{c_i \wedge_i (t) + C_i^P(t) \wedge_i (t)\}, \tag{4.4}$$

where c_{s,k_j} is the cost of each system failure caused by the failure of component k_j, c_i is the cost of the failure of component k, $C_i^P(t)$ is the cost of PM on other components when component i fails, $\wedge U_{k_j}(t)$ is the expected number of failures for the minimum cut set U_{k_j}, and U_{k_j} is a first-order cut set.

Because $\wedge_i(t) = -\ln(R_i(t))$, the definition of the importance measure in Eq. (4.4) can be re-written as

$$\begin{aligned} I'^{C}_{j|i} &= H_{j|i} \frac{-\partial\, C'(t)}{\partial R_i(t)} \\ &= -\, H_{j|i} \frac{\partial\, C^{'}(t)}{\partial \wedge_i(t)} \frac{\wedge_i(t)}{\partial R_i(t)} \\ &= H_{j|i} \frac{1}{R_i(t)} \frac{\partial\, C^{'}(t)}{\partial \wedge_i(t)}. \end{aligned} \tag{4.5}$$

In this way it is also possible to consider the extent to which PM affect system costs.

4.1.2.2 Cost-Based Component Maintenance Importance (CCMI) in Multistate Systems

The concept of CCMI can be applied in multistate systems. Suppose that state K is the threshold system state: if the system state is below K, the system is in a state needing maintenance in order to keep the system at a level of performance. Similarly, state K_i is the threshold state of component i. As long as the state of component i is below K_i, the degradation of the component can be detected and restored immediately.

Definition 4.2 If component i has degraded into a state below K_i, the CCMI of component j is defined by

$$I^C_{j|i}(t) = -H_{j|i} I^c_{j|i}(t). \tag{4.6}$$

The symbol $(< K)_i$ represents that the state of component i degrades into a state below state K_i, $\chi\{.\}$ is an indicator function, $I^c_{j|i}(t)$ is the cost-based importance of component j, given that component i's state is below K_i.

Under the following two scenarios, we give the corresponding expressions of $I^c_{j|i}(t)$ to analyse the effect of component j on the total maintenance cost when component i in a multistate system is maintained.

Scenario 1: If the observed state of component i is $(K_o)_i$ and other component states cannot be observed, then the threshold state K_j of component j $(j \neq i)$ can be used to represent the degradation of component j. The repair cost of a multi-state component is the expected cost per state if the state is below its threshold.

For multi-state components, when the observed state of component i is $(K_o)_i$, similarly to Eq. (4.1), we have

$$C(t) = \sum_{i=1}^{n} \left\{ \left\{ \begin{array}{c} c_{s,i^{(<K)_i}} \Pr\left[\phi((<K)_i, X(t)) < K\right] \\ + c_{i^{(<K)_i}} + C^P_{i^{(<K)_i}}(t) \end{array} \right\} \Pr\left[x_i = (<K)_i\right] \right\}, \quad (4.7)$$

and

$$\begin{aligned} C^P_{i^{(<K)_i}}(t) &= H_{s,i} \sum_{j=1, i \neq j}^{n} c_{p_j} \Pr\left[x_j \geq K_j\right] \\ &+ (1 - H_{s,i}) \sum_{z=1}^{m} c_{p_{j_z}} \Pr\left[x_{j_z} \geq K_{j_z}\right] \Pr\left[\begin{array}{c} \phi((<K)_i, (<K)_{j_1}, \ldots, \\ (<K)_{j_{z-1}}, (\geq K)_{i, j_1, j_2, \ldots, j_{z-1}}) \geq K \end{array} \right]. \end{aligned} \quad (4.8)$$

Assume that a PM on an item can restore the item to its as-good-as new status. Thus, $c_{i^{(<K)_i}}$ is the cost of restoring component i from state $(<K)_i$ to state M_i, and $c_{s,i^{(<K)_i}}$ is the cost per system failure caused by the state of component i below K_i. c_{p_j} is the cost of PM of component j until it is in the perfect state. Based on Eq. (4.8), we have

$$I^c_{j|i} = -\frac{\partial C((K_0)_i, X(t))}{\partial \rho_{jK_j}(t)}. \quad (4.9)$$

Equation (4.9) shows the impact on component j, which is under PM, on the total system cost when component i fails.

Scenario 2: Because the states of all components can be observed, the cost of PM changes over component states. More precisely, the maintenance cost on component j is the cost of restoring the component j from state $(K_o)_j$ to state M_j.

Based on Eq. (4.8), we have

$$\begin{aligned} C^P_{i^{(<K)_i}}(t) &= H_{s,i} \sum_{j=1, i \neq j}^{n} c_{p_{jK_0-M}} \Pr\left[x_j \geq K_j\right] \\ &+ (1 - H_{s,i}) \left(\begin{array}{c} \sum_{z=1}^{m} c_{p_{j_z K_0-M}} \Pr\left[x_{j_z} \geq K_{j_z}\right] \\ \Pr\left[\phi((<K)_i, (<K)_{j_1}, \ldots, (<K)_{j_{z-1}}, (\geq K)_{i, j_1, j_2, \ldots, j_{z-1}}) \geq K\right] \end{array} \right), \end{aligned} \quad (4.10)$$

where $c_{p_{jK_0-M}}$ represents the cost of restoring component j from state $(K_o)_j$ to state M_j. Then we can use Eq. (4.11) to analyse the impact on PM component j

on the total system cost when component i fails.

$$I^{c}_{j|i} = -\frac{\partial C((K_0)_i, X(t))}{\partial \rho_{j(K_0)_j}(t)}. \tag{4.11}$$

4.1.2.3 Discussion on Maintenance Optimisation

In the preceding sections, we assume that component j is maintained to state M_j. Obviously, in reality, one may have more options. The effectiveness of the repair on a failed component can be better-than-perfect, perfect, imperfect, minimal, or worse-than-minimal.

Under Scenario 2, let us assume that the maintenance effectiveness on component i is minimal, better-than-perfect, or perfect. Based on Eqs. (4.7) and (4.8), we obtain the following cost function, considering the different maintenance quality level.

$$\begin{aligned} C(t) = &\sum_{i=1}^{n} c_{s,i^{K_0-K_{l1}}} \Pr\left[\phi((K_0)_i, X(t)) < K\right] \Pr\left[x_i = (K_0)_i\right] \\ &+ \left[c_{i^{K_0-K_{l1}}} + C^P_{i^{K_0-K_{l1}}}(t)\right] \Pr\left[x_i = (K_0)_i\right], \end{aligned} \tag{4.12}$$

$$\begin{aligned} C^P_{i^{K_0-K_l}}(t) = &H_{s,i} \sum_{j=1, i\neq j}^{n} c_{p_j{}^{K_0-K_{l2}}} \Pr\left[x_j \geq K_j\right] \\ &+ (1 - H_{s,i}) \left(\frac{\sum_{z=1}^{m} c_{p_{j_z}{}^{K_0-K_{l2}}} \Pr\left[x_{j_z} \geq K_{j_z}\right]}{\Pr\left[\phi((< K)_i, (< K)_{j_1}, \ldots, (< K)_{j_{z-1}}, (\geq K)_{i, j_1, j_2, \ldots, j_{z-1}}) \geq K\right]} \right), \end{aligned} \tag{4.13}$$

in which $l_1 \in \{1, 2, 3\}$, $l_2 \in \{2, 3\}$. Assume all component states corresponding to the performance can be observed and the observed states of components i and j are $(K_o)_i$ and $(K_o)_j$, respectively. $c_{s,i^{K_0-K_{l1}}}$ is the cost per system failure incurred due to maintaining of failed component i from state $(K_o)_i$ to state $(K_{l_1})_i$. The $c_{i^{K_0-K_{l1}}}$ is the cost of maintaining component i from state $(K_o)_i$ to state $(K_{l_1})_i$.

Then a new question arises: Which state of the component should be restored to its perfect state? Here we adopt the method in Wu and Coolen [21] and define a cost-based improvement potential (CIP) in a binary system

$$I^P_i(t) = C(1_i, X(t)) - C(X(t)). \tag{4.14}$$

$I^P_i(t)$ is the difference between the maintenance cost with a perfect component i and the maintenance cost with the actual component i.

A cost-based improvement potential in a multistate system is given by

$$I^P_{iK_{0}-K_{l1}}(t) = C\left((K_{l_1})_i, X(t)\right) - C\left((K_0)_i, X(t)\right), \quad (4.15)$$

where $C((K_{l_1})_i, X(t))$ is the cost of maintaining the system when the state of component i is improved to be $(K_{l_1})_i$ and the states of the other components remain unchanged, and $C((K_0)_i, X(t))$ is the cost of maintaining the system when the observed state of component i is $(K_o)_i$, i.e. the cost $C(t)$ are shown in Eq. (4.12), $I^P_{iK_0-K_{l1}}(t)$ is the difference between the maintenance cost with component i at state $(K_{l_1})_i$ and the maintenance cost with component i in state $(K_o)_i$.

We can therefore define the impact of system maintenance costs due to the change in the state of component i.

$$I^S_{ik_{l_1}}(t) = \frac{I^P_{iK_0-K_{l1}}(t)}{K_0 - K_{l_1}}. \quad (4.16)$$

We should choose the value of l_1 corresponding to the largest $I^S_{ik_{l_1}}(t)$. K_{l_1} is the best choice for maintaining component i.

4.1.2.4 Maintenance Policies

This section defines two maintenance policies: maintenance policies A and B, as shown below.

Maintenance policy A. Once the state of a component degrades into a state below its threshold state, it will be immediately observed and then maintained. Under this policy, if the failed component is a critical component, the system stops working. PM can be performed on all other components. If the failed component is a non-critical component, the system is still working. PM can be performed on non-critical components, but not on critical ones.

Definition 4.3 Under maintenance policy A, if the state of component i is below its threshold state K_i, then, the CCMI of component j is defined by

$$I^C_{j|i}(t) = -H_{j|i} I^c_{j|i}(t). \quad (4.17)$$

If the system structure function $\Phi(.)$ reduces to a state that is below its threshold state K, i.e., $\Phi((< K)_i, X(t)) < K$, it is obvious that the state of the corresponding component is lower than the threshold, and the component is a critical component. At this time, the system will stop working. Thus, the PM can be performed on all other components, $j \in \{1, \ldots, i-1, i+1, \ldots, n\}$. When component i is non-critical, the system will not fail and still keep working, i.e., $\Phi((< K)_i, X(t)) \geq K$. Hence, the PM can be performed on the non-critical components, $j \in \left\{j | \Phi\left((< K)_i, (< K)_j, X(t)\right) \geq K\right\}$.

If Scenarios 1 and 2 use maintenance policy A, we only need to substitute Eqs. (4.9) and (4.11) into Eq. (4.17), respectively.

For specific applications, when component i fails, we can choose the component with the largest $I^C_{j|i}(t)$ as the first component to perform PM because this component can minimise the total cost. Then, according to the rankings of $I^C_{j|i}(t)$, we can select components in turn for PM. In the following, we will discuss this question in detail.

Maintenance policy B. Suppose that the system state degrades into a state below its threshold state K. During maintenance activities, the corresponding components can be located. These components may contain some critical components or some non-critical components. The set of components $i_1, i_2, \ldots, i_{m_i}$ is a cut set of the system. This means that downgrading the components of the collection below the threshold state will cause the system state to degrade below its threshold state K. Therefore, under maintenance policy B, when components $i_1, i_2, \ldots, i_{m_i}$ are all degraded into states below the threshold K, they are maintained and the system must stop working. In this situation, PM can be performed on all other components Generally, under policy B, a system fails because at least one minimal cut set failed, and the PM can be performed on all other components. Assume that there are n_0 minimal cut sets in the system, then the cost incurred within time interval $(0, t)$ is composed of three parts: one on cost of repairing the system upon failures of the n_0 minimal cut sets, one on cost of repairing each component in the failed minimal cut sets, and one on cost of all components in the system except the component in the failed minimal cut sets. So, we have

$$C(t)=(c_{s,i}+\sum_{j=1}^{m_i} c_{i_j}+\sum_{z=1}^{n-m_i} c_{p_z})\Pr\left[\phi((K_0)_{i_1}, (K_0)_{i_2}, \ldots, (K_0)_{i_{m_i}}, X(t)) < K\right], \tag{4.18}$$

where $c_{s,i}$ is the expected cost per system failure, caused by the failure of the ith minimum cut set, and c_{i_j} is the expected cost per failure of component i_j.
Under Scenario 1, the states of the components $i_1, i_2, \ldots, i_{m_i}$ are $(K_o)_{i_1}$, $(K_o)_{i_2}, \ldots, (K_o)_{i_{m_i}}$, which can be observed, respectively. But the state of other components cannot be observed, then we have

$$\begin{aligned} I^C_{j|i_1,i_2,\ldots,i_{m_i}}(t) &= I^c_{j|i_1,i_2,\ldots,i_{m_i}}(t) \\ &= \frac{-\partial C((K_0)_{i_1}, (K_0)_{i_2}, \ldots, (K_0)_{i_{m_i}}, X(t))}{\partial \rho_{jK_j}(t)}. \end{aligned} \tag{4.19}$$

Under Scenario 2, the states of the components $i_1, i_2, \ldots, i_{m_i}$ are $(K_o)_{i_1}$, $(K_o)_{i_2}, \ldots, (K_o)_{i_{m_i}}$, which can be observed, respectively. But the state of other components can be observed as well, then we have

$$\begin{aligned} I^C_{j|i_1,i_2,\ldots,i_{m_i}}(t) &= I^c_{j|i_1,i_2,\ldots,i_{m_i}}(t) \\ &= \frac{-\partial C((K_0)_{i_1}, (K_0)_{i_2}, \ldots, (K_0)_{i_{m_i}}, X(t))}{\partial \rho_{j(K_0)_j}(t)}. \end{aligned} \tag{4.20}$$

For specific applications, if components $\{i_1, i_2, \ldots, i_{m_i}\}$ degrade into below their threshold state, we can choose components j with the largest $I^C_{j|i_1,i_2,\ldots,i_{m_i}}(t)$ as the first component for PM because this component can minimise the loss of the total cost. Then, according to the ranking of $I^C_{j|i_1,i_2,\ldots,i_{m_i}}(t)$, we can select components in turn for PM.

4.1.3 A Cost-Based IIM

4.1.3.1 The Definition of Cost-Based IIM

When a component is failed at time t, assuming the repair time of the component is Δt, we analyse how the system reliability changes in Δt.

$$\begin{aligned} & R(t+\Delta t) - R(t) \\ & = \begin{pmatrix} f(R_1(t+\Delta t), R_2(t+\Delta t), \ldots, R_n(t+\Delta t)) \\ -f(R_1(t), R_2(t), \ldots, R_n(t)) \end{pmatrix} \\ & = \begin{pmatrix} f(R_1(t+\Delta t), R_2(t+\Delta t), \ldots, R_n(t+\Delta t)) \\ -f(R_1(t), R_2(t+\Delta t), \ldots, R_n(t+\Delta t)) \\ +f(R_1(t), R_2(t+\Delta t), \ldots, R_n(t+\Delta t)) \\ -f(R_1(t), R_2(t), R_3(t+\Delta t) \ldots, R_n(t+\Delta t)) \ldots \\ +f(R_1(t), R_2(t), \ldots, R_{n-1}(t), R_n(t+\Delta t)) \\ -f(R_1(t), R_2(t), \ldots, R_n(t)). \end{pmatrix}. \end{aligned} \tag{4.21}$$

Then the change of the system reliability in a unit of time is

$$\begin{aligned} & \frac{R(t+\Delta t)-R(t)}{\Delta t} \\ & = \frac{\begin{pmatrix} f(R_1(t+\Delta t), R_2(t+\Delta t), \ldots, R_n(t+\Delta t)) \\ -f(R_1(t), R_2(t+\Delta t), \ldots, R_n(t+\Delta t)) \end{pmatrix}}{\Delta t} \\ & + \frac{\begin{pmatrix} f(R_1(t), R_2(t+\Delta t), \ldots, R_n(t+\Delta t)) \\ -f(R_1(t), R_2(t), R_3(t+\Delta t) \ldots, R_n(t+\Delta t)) \end{pmatrix}}{\Delta t} \\ & + \cdots + \frac{\begin{pmatrix} f(R_1(t), R_2(t), \ldots, R_{n-1}(t), R_n(t+\Delta t)) \\ -f(R_1(t), R_2(t), \ldots, R_n(t)) \end{pmatrix}}{\Delta t} \\ & = \frac{R_1(t+\Delta t)-R_1(t)}{\Delta t} \frac{\begin{pmatrix} f(R_1(t+\Delta t), R_2(t+\Delta t), \ldots, R_n(t+\Delta t)) \\ -f(R_1(t), R_2(t+\Delta t), \ldots, R_n(t+\Delta t)) \end{pmatrix}}{R_1(t+\Delta t)-R_1(t)} \\ & + \frac{R_2(t+\Delta t)-R_2(t)}{\Delta t} \frac{\begin{pmatrix} f(R_1(t), R_2(t+\Delta t), \ldots, R_n(t+\Delta t)) \\ -f(R_1(t), R_2(t), R_3(t+\Delta t) \ldots, R_n(t+\Delta t)) \end{pmatrix}}{R_2(t+\Delta t)-R_2(t)} \\ & + \cdots + \frac{R_n(t+\Delta t)-R_n(t)}{\Delta t} \frac{\begin{pmatrix} f(R_1(t), R_2(t), \ldots, R_{n-1}(t), R_n(t+\Delta t)) \\ -f(R_1(t), R_2(t), \ldots, R_n(t)) \end{pmatrix}}{R_n(t+\Delta t)-R_n(t)}. \end{aligned} \tag{4.22}$$

In the repair time Δt of a failed component (component i, for example), the system reliability may approximately reduce by $\Delta t I_i^{\text{im}}(t)$ due to the failure of component i, based on the Taylor expansion of $R(t)$. Then the component that brings the maximum decrement of system reliability should be firstly performed on the PM, i.e., the component whose IIM is the biggest.

The cost due to the failure of component i includes the cost of maintaining the system, which is caused by the failure of component i, and the cost of maintaining component i.

Then the total maintenance cost due to the failure of component i is given by

$$C(i,t) = C_{s,i}(t) + C_i(t). \tag{4.23}$$

in which, $C_{s,i}(t)$ is the cost of maintaining the system, caused by the failure of component i, and $C_i(t)$ is the cost of maintaining component i.

In a system consisting of n components, when a component fails, the component which should be firstly performed on the PM is required to satisfy two conditions: $\min\limits_{i=1,2,\ldots,n}\{C(i,t)\}$ and $\max\limits_{i=1,2,\ldots,n}\{I_i^{\text{im}}(t)\}$.

Definition 4.4 The cost-based IIM of component i is defined by

$$I_i^{cim}(t) = \frac{C(i,t)}{I_i^{\text{im}}(t)}. \tag{4.24}$$

From Definition 4.4, the component with the minimum cost-based IIM $(\min\limits_{i=1,2,\ldots,n}\{I_i^{cim}(t)\})$ should be firstly performed on the PM.

When a critical component has failed, the system fails. Then $I_i^{cim}(t)$ can be performed on all other components. When a non-critical component has failed, the system does not fail. In order to ensure system operation smoothly, $I_i^{cim}(t)$ can be performed on all other non-critical components.

Definition 4.5 The cost-based IIM of a group of components $i_1, i_2 \ \ldots, \ i_m$ is defined by

$$I_{i_1,i_2,\ldots,i_m}^{cim}(t) = \frac{C(i_1,t) + C(i_2,t) + \cdots + C(i_m,t)}{I_{i_1,i_2,\ldots,i_m}^{im}(t)}. \tag{4.25}$$

Based Definition 4.5, multiple components can be identified for PM, and the system reliability can therefore be improved and the maintenance cost can be reduced.

The number of components for PM is related to the constraint of the total maintenance cost. When a critical component fails, PM can be performed on all other components. When a non-critical component fails, PM can be performed on all other non-critical components. We assume the total maintenance cost is C, and the order of the identified components performed on PM is components $1, 2, \ldots, n$. Then the number of components for PM is $N = \{N | \sum_{i=1}^{n} C(i,t) \le C \le \sum_{i=1}^{N+1} C(i,t)\}$.

In the following, we investigate some characteristics of cost-based IIM and show how the cost-based IIM works.

Example 4.1 In a series system consisting of n components, the system reliability is

$$\begin{aligned} R(t) &= \prod_{i=1}^{n} R_i(t) \\ &= \prod_{i=1}^{n} \exp(-\int_0^t \lambda_i(u)du) \\ &= \exp(-\sum_{i=1}^{n} \int_0^t \lambda_i(u)du). \end{aligned} \tag{4.26}$$

Assume the maintenance costs of all components are the same, $C(i, t) = C(t)$, then the cost-based IIM of component i is

$$\begin{aligned} I_i^{cim}(t) &= \frac{C(t)}{\lambda_i(t)\exp(-\int_0^t \lambda_i(u)du)\frac{\partial R(t)}{\partial R_i(t)}} \\ &= \frac{C(t)}{\lambda_i(t)\exp(-\sum_{k=1}^{n}\int_0^t \lambda_k(u)du)}. \end{aligned} \tag{4.27}$$

Comparing the cost-based IIM of two different components i and j,

$$\begin{aligned} \frac{I_i^{cim}(t)}{I_j^{cim}(t)} &= \frac{\lambda_j(t)\exp(-\sum_{k=1}^{n}\int_0^t \lambda_k(u)du)}{\lambda_i(t)\exp(-\sum_{k=1}^{n}\int_0^t \lambda_k(u)du)} \\ &= \frac{\lambda_j(t)}{\lambda_i(t)}. \end{aligned} \tag{4.28}$$

So we have

$$\begin{aligned} & I_i^{cim}(t) < I_j^{cim}(t) \\ & \Leftrightarrow \frac{I_i^{cim}(t)}{I_j^{cim}(t)} < 1 \\ & \Leftrightarrow \frac{\lambda_j(t)}{\lambda_i(t)} < 1 \\ & \Leftrightarrow \lambda_j(t) < \lambda_i(t). \end{aligned} \tag{4.29}$$

Thus, in a series system, if $\lambda_i(t) < \lambda_j(t)$, then $I_i^{cim}(t) < I_j^{cim}(t)$. The component with the minimal cost-based IIM should have the top priority for PM. That is, in a series system, components with larger failure rates should have higher priority for PM.

When the lifetime distribution function of each component follows the exponential distribution, the Gamma distribution, or the Weibull distribution, we can obtain the following inferences, respectively.

- In a series system, when the lifetime distribution function of component i follows the Gamma distribution with shape parameter α_i and scale parameter β_i, $\frac{1}{\lambda_i(t)} = \int_0^\infty (1+\frac{u}{t})^{\alpha_i-1}e^{-\beta_i u}du$. So

$$\begin{aligned}\lambda_i(t) > \lambda_j(t) &\Leftrightarrow \frac{\lambda_i(t)}{\lambda_j(t)} > 1\\ &\Leftrightarrow \frac{\int_0^\infty (1+\frac{u}{t})^{\alpha_j-1}e^{-\beta_j u}du}{\int_0^\infty (1+\frac{u}{t})^{\alpha_i-1}e^{-\beta_i u}du} > 1\\ &\Leftrightarrow \int_0^\infty ((1+\frac{u}{t})^{\alpha_j-1}e^{-\beta_j u} - (1+\frac{u}{t})^{\alpha_i-1}e^{-\beta_i u})du > 0. \end{aligned} \tag{4.30}$$

When $\alpha_i < \alpha_j, \beta_i > \beta_j$, we have $\lambda_i(t) > \lambda_j(t)$. So, we can obtain that if $\alpha_i < \alpha_j, \beta_i > \beta_j$, then $I_i^{cim}(t) < I_j^{cim}(t)$.

- In a series system, when the lifetime distribution function of component i follows the Weibull distribution with shape parameter α_i and scale parameter β_i, $\lambda_i(t) = \beta_i\alpha_i(\beta_i t)^{\alpha_i-1}$. We can obtain

$$\begin{aligned}\lambda_i(t) > \lambda_j(t) &\Leftrightarrow \frac{\lambda_i(t)}{\lambda_j(t)} > 1\\ &\Leftrightarrow \frac{\beta_i\alpha_i(\beta_i t)^{\alpha_i-1}}{\beta_j\alpha_j(\beta_j t)^{\alpha_j-1}} > 1\\ &\Leftrightarrow \frac{\beta_i\alpha_i(\beta_i)^{\alpha_i-1}}{\beta_j\alpha_j(\beta_j)^{\alpha_j-1}} t^{\alpha_i-\alpha_j-2} > 1\\ &\Leftrightarrow \frac{\alpha_i}{\alpha_j}\frac{{\beta_i}^{\alpha_i-\alpha_j}}{(\beta_j/\beta_i)^{\alpha_j}} t^{\alpha_i-\alpha_j-2} > 1. \end{aligned} \tag{4.31}$$

When $t > 1, \alpha_i > \alpha_j+2$, we obtain $t^{\alpha_i-\alpha_j-2} > 1$, $\frac{\alpha_i}{\alpha_j} > 1$. When $\beta_i > \beta_j > 1$, we have ${\beta_i}^{\alpha_i-\alpha_j} > 1$, $(\beta_j/\beta_i)^{\alpha_j} < 1$, so $\frac{{\beta_i}^{\alpha_i-\alpha_j}}{(\beta_j/\beta_i)^{\alpha_j}} > 1$. Then when $t > 1, \alpha_i > \alpha_j+2, \beta_i > \beta_j > 1$, we have $\lambda_i(t) > \lambda_j(t)$. So, we can obtain that if $t > 1, \alpha_i > \alpha_j+2$, $\beta_i > \beta_j > 1$, then $I_i^{cim}(t) < I_j^{cim}(t)$.

From the above inferences, based on the relationships between the failure rates, shape parameters and scale parameters of different components, we can easily determine which component should be selected for PM.

Example 4.2 In a parallel system consisting of n components, the system reliability is $R(t) = 1 - \prod_{i=1}^n (1 - R_i(t)) = 1 - \prod_{i=1}^n \left(1 - \exp(-\int_0^t \lambda_i(u)du)\right)$.

Assume the maintenance costs of all components are the same, $C(i,t) = C(t)$, then the cost-based IIM of component i is

$$I_i^{cim}(t) = \frac{C(t)}{\lambda_i(t)\exp(-\int_0^t \lambda_i(u)du)\frac{\partial R(t)}{\partial R_i(t)}} \times \frac{C(t)}{\lambda_i(t)\exp(-\int_0^t \lambda_i(u)du)\prod_{k=1,k\neq i}^{n}\left(1-\exp(-\int_0^t \lambda_k(u)du)\right)}. \quad (4.32)$$

Then we have

$$\frac{I_i^{cim}(t)}{I_j^{cim}(t)} = \frac{\lambda_j(t)\exp(-\int_0^t \lambda_j(u)du)\prod_{k=1,k\neq j}^{n}\left(1-\exp(-\int_0^t \lambda_k(u)du)\right)}{\lambda_i(t)\exp(-\int_0^t \lambda_i(u)du)\prod_{k=1,k\neq i}^{n}\left(1-\exp(-\int_0^t \lambda_k(u)du)\right)} = \frac{\lambda_j(t)\exp(-\int_0^t \lambda_j(u)du)\left(1-\exp(-\int_0^t \lambda_i(u)du)\right)}{\lambda_i(t)\exp(-\int_0^t \lambda_i(u)du)\left(1-\exp(-\int_0^t \lambda_j(u)du)\right)}. \quad (4.33)$$

So, we can obtain

$$\begin{aligned} I_i^{cim}(t) < I_j^{cim}(t) &\Leftrightarrow \frac{I_i^{cim}(t)}{I_j^{cim}(t)} < 1 \\ &\Leftrightarrow \frac{\lambda_j(t)\exp\left(-\int_0^t \lambda_j udu\right)\left(1-\exp\left(-\int_0^t \lambda_i udu\right)\right)}{\lambda_i(t)\exp\left(-\int_0^t \lambda_i udu\right)\left(1-\exp\left(-\int_0^t \lambda_j udu\right)\right)} < 1 \\ &\Leftrightarrow (\lambda_i(t)-\lambda_j(t))\exp\left(-\int_0^t \lambda_i udu\right)\exp\left(-\int_0^t \lambda_j udu\right) \\ &< \lambda_i(t)\exp\left(-\int_0^t \lambda_i udu\right) - \lambda_j(t)\exp\left(-\int_0^t \lambda_j udu\right). \end{aligned} \quad (4.34)$$

When $\lambda_j(t) > \lambda_i(t)$, we have

$$\left(\lambda_i(t)-\lambda_j(t)\right)\exp\left(-\int_0^t \lambda_i udu\right)\exp\left(-\int_0^t \lambda_j udu\right) < 0. \quad (4.35)$$

When $\lambda_j(t) > \lambda_i(t)$, and $\int_0^t \left(\lambda_j u - \lambda_i(u)\right)du > \ln\lambda_j(t) - \ln\lambda_i(t)$, we have

$$\begin{aligned} &\int_0^t \left(\lambda_j u - \lambda_i u\right)du > \ln\lambda_j(t) - \ln\lambda_i(t) \\ &\Leftrightarrow \int_0^t \left(\lambda_j u - \lambda_i u\right)du > \ln\frac{\lambda_j(t)}{\lambda_i(t)} \\ &\Leftrightarrow \exp\left(\int_0^t \left(\lambda_j u - \lambda_i u\right)du\right) > \frac{\lambda_j(t)}{\lambda_i(t)} \end{aligned}$$

$$\Leftrightarrow \frac{\lambda_i(t)\exp\left(-\int_0^t \lambda_i u du\right)}{\lambda_j(t)\exp\left(-\int_0^t \lambda_j u du\right)} > 1$$

$$\Leftrightarrow \lambda_i(t)\exp\left(-\int_0^t \lambda_i u du\right) - \lambda_j(t)\exp\left(-\int_0^t \lambda_j u du\right) > 0. \tag{4.36}$$

Thus, in a parallel system, if $\lambda_j(t) > \lambda_i(t)$, and $\int_0^t \left(\lambda_j u - \lambda_i(u)\right) du > \ln\lambda_j(t) - \ln\lambda_i(t)$, then $I_i^{cim}(t) < I_j^{cim}(t)$.

In series and parallel systems, we can obtain the order of the cost-based IIM values of all components based on the above discussed relationships among the component failure rates, so as to arrange the PM for maintainers.

4.1.3.2 Optimisation of PM Policy

Age-dependent PM policy Under this maintenance policy, component i is preventively maintained at predetermined age T, or repaired at failure, whichever comes first. When PM is performed on component i, component i is still working with probability $R_i(t)$. When component i is being repaired, it may fail with probability $F_i(t) = 1 - R_i(t)$. So in a maintenance cycle, the expected operational time of component i is

$$P_i^{EOT}(T) = \int_0^T t dF_i(t) + \int_T^\infty T dF_i(t)$$
$$= \int_0^T R_i(t)dt. \tag{4.37}$$

The expected maintenance time is

$$P_i^{EMT}(T) = \tau_{i,f} F_i(T) + \tau_{i,p} R_i(T),$$

in which $\tau_{i,f}$ is the repair time, and $\tau_{i,p}$ is the PM time. So the expected cycle length is

$$P_i^{ECL}(T) = P_i^{EOT}(T) + P_i^{EMT}(T) = \int_0^T R_i(t)dt + \tau_{i,f} F_i(T) + \tau_{i,p} R_i(T). \tag{4.38}$$

The expected cycle cost is $P_i^{ECC}(C) = c_{i,f} F_i(T) + c_{i,p} R_i(T)$, in which $c_{i,f}$ is the repair cost, and $c_{i,p}$ is the PM cost. Then the average maintenance cost of component i in a maintenance cycle is

$$P_i^{CIM}(C) = \frac{P_i^{ECC}(C)}{P_i^{ECL}(T)}$$
$$= \frac{c_{i,f}F_i(T) + c_{i,p}R_i(T)}{\int_0^T R_i(t)dt + \tau_{i,f}F_i(T) + \tau_{i,p}R_i(T)}. \quad (4.39)$$

In a maintenance cycle, component i is preventively maintained, or repaired upon failures, so the failure rate of component i changes with the maintenance action. The failure rate of component i is approximated by the reciprocal of the expected operational time of component i, i.e.,

$$\lambda_i(T) = \frac{1}{P_i^{EOT}(T)} = \frac{1}{\int_0^T R_i(t)dt}. \quad (4.40)$$

The reliability of component i is approximated by the ratio of the expected operational time and expected cycle length, i.e.

$$R_i(T) = \frac{P_i^{EOT}(T)}{P_i^{ECL}(T)}$$
$$= \frac{\int_0^T R_i(t)dt}{\int_0^T R_i(t)dt + \tau_{i,f}F_i(T) + \tau_{i,p}R_i(T)}. \quad (4.41)$$

So based on Eqs. (4.40) and (4.41), at the predetermined age T, the IIM of component i is

$$I_i^{\text{im}}(T) = \lambda_i(T)R_i(T)\frac{\partial R(T)}{\partial R_i(T)}$$
$$= \frac{1}{P_i^{EOT}(T)}\frac{P_i^{EOT}(T)}{P_i^{ECL}(T)}\frac{\partial R(T)}{\partial R_i(T)}$$
$$= \frac{1}{P_i^{ECL}(T)}\frac{\partial R(T)}{\partial R_i(T)}. \quad (4.42)$$

At the predetermined age T, the cost-based IIM of component i is

$$I_i^{cim}(T) = \frac{P_i^{CIM}(T)}{I_i^{\text{im}}(T)} = \frac{\frac{P_i^{ECC}(C)}{P_i^{ECL}(T)}}{\frac{1}{P_i^{ECL}(T)}\frac{\partial R(T)}{\partial R_i(T)}} = \frac{P_i^{ECC}(C)}{\frac{\partial R(T)}{\partial R_i(T)}}. \quad (4.43)$$

So, under the age-dependent PM policy, we can identify which component should be selected for PM at predetermined age T.

Block maintenance policy Under this maintenance policy, component i is preventively maintained at predetermined age T, $2T$, $3T$ ... or repaired at failure, whichever comes first. The block maintenance policy is usually used on a system consisting of a group of components of low capitalisation value, such as individual light fixtures, batch of fire extinguishers, batch of fire alarm initiating devices, and so on. The ages of the components are not observed and only their failures are known, all components may be replaced periodically independently of their ages in use. Assume a group of components $i_1, i_2 \ \ldots, \ i_m$ is preventively maintained or repaired simultaneously. We can use Definition 4.5 (the cost-based im of a group of components $i_1, i_2 \ \ldots, \ i_m$) to identify which group of components should be selected for PM.

According to the description of age-dependent PM policy, we have

$$I_{i_1}^{im}(t) + I_{i_2}^{im}(t) + \cdots + I_{i_m}^{im}(t) = \frac{1}{P_{i_1}^{ECL}(T)} \frac{\partial R(T)}{\partial R_{i_1}(T)} + \frac{1}{P_{i_2}^{ECL}(T)} \frac{\partial R(T)}{\partial R_{i_2}(T)} + \cdots + \frac{1}{P_{i_m}^{ECL}(T)} \frac{\partial R(T)}{\partial R_{i_m}(T)}. \tag{4.44}$$

Based on Eq. (4.43), the cost-based IIM of a group of components $i_1, i_2, \ldots, i_m$ is

$$I_{i_1,i_2,\ldots,i_m}^{cim}(T) = \frac{P_{i_1}^{AMC}(C) + P_{i_2}^{AMC}(C) + \cdots + P_{i_m}^{AMC}(C)}{\frac{1}{P_{i_1}^{ECL}(T)} \frac{\partial R(T)}{\partial R_{i_1}(T)} + \frac{1}{P_{i_2}^{ECL}(T)} \frac{\partial R(T)}{\partial R_{i_2}(T)} + \cdots + \frac{1}{P_{i_m}^{ECL}(T)} \frac{\partial R(T)}{\partial R_{i_m}(T)}}. \tag{4.45}$$

From Eq. (4.45), under the block maintenance policy, multi-components can be identified for PM.

4.2 Cost-Based Joint Importance Measures for Optimisation of PM Policies

Reducing the economic loss caused by a component failure and increasing the system availability through effective maintenance activities are important in reliability engineering and can be achieved through proper maintenance. However, improperly improving system reliability may cause a negative consequence. Existing literature lacks a discussion of maintenance policies on the relationship between the alteration of system reliability and their implications on the costs from the perspective of the system lifetime, which will be addressed below.

The continual development of engineered systems and the increasing reliance on equipment have led to the increasing importance of improving component lifetime and reducing maintenance costs. In order to better maintain systems [11], many researchers have conducted related studies. Iscioglu [12] analysed the remaining life function of a polymorphic system and evaluated the situation when the life cycles are independent and dependent on each other. Bohlooli-Zefreh et al. [4] further applied a cost function to obtain the optimal replacement policy for the system based on their proposed failure model. Levitin et al. [14] studied a method to evaluate the expected unfinished part of the task and the survivability of the system experiencing internal failures and external shocks. Tan et al. [20] proposed an index for the remaining life that considered the degree of damage in the time period. The index is used to evaluate maintenance policies to improve system lifetime.

Resource constraint is also a challenge for many organisations. Issues such as optimising resource allocation and maintenance optimisation have been discussed [16]. Yuan et al. [28] proposed a potential cost index, which was used to exclude measures that exceed budget. At the same time, they constructed a knapsack problem that is to solve the corresponding measures to maximise the system lifetime under the limited budget. Jafary et al. [13] proposed a method using explicit correlation parameters to characterise related component failures, and then gave an optimal maintenance policy to reduce system failures. Levitin et al. [15] studied an optimal multiple replacement and maintenance scheduling considering the resource limitation. One needs to conduct an in-depth investigation on planning issues on the system performance optimisation from the perspective of selection of maintenance components under resource constraints.

4.2.1 Different Cost Analysis on System Lifetime Change

Let $F = F(\boldsymbol{q}(t)) = F(q_1(t), q_2(t), \cdots q_n(t))$ be the system lifetime distribution, and $q_i(t)$ be the lifetime distribution of component i. Then $F(0_i, \boldsymbol{X}(t)) - F(1_i, \boldsymbol{X}(t))$ quantifies the difference between the lifetime distributions of restoring component i from a failed state to a normal state.

The joint failure importance can then be given by

$$I_{i,k}^{f} = F(0_i, 0_k, \boldsymbol{X}(t)) + F(1_i, 1_k, \boldsymbol{X}(t)) - F(1_i, 0_k, \boldsymbol{X}(t)) - F(0_i, 1_k, \boldsymbol{X}(t)). \tag{4.46}$$

If component i fails, the system lifetime distribution $F(q)$ becomes $F(0_i, \boldsymbol{X}(t))$. In the case of component failure, the probability density function of the system is

$$\begin{aligned}\frac{dF(0_i, \boldsymbol{X}(t))}{dt} &= \frac{dF_k(t)}{dt}\frac{dF(0_i, \boldsymbol{X}(t))}{dF_k(t)} \\ &= \frac{dF_k(t)}{dt}\left(F(0_i, 0_k, \boldsymbol{X}(t)) - F(0_i, 1_k, \boldsymbol{X}(t))\right) \\ &= \lambda_k(t)R_k(t)\left(F(0_i, 0_k, \boldsymbol{X}(t)) - F(0_i, 1_k, \boldsymbol{X}(t))\right),\end{aligned} \tag{4.47}$$

where $\lambda_k(t) = \frac{dF_k(t)}{dt}$. Let $I_k(t)_{X_i(t)=0} = \lambda_k(t)R_k(t)\left[F(0_i, 0_k, \boldsymbol{X}(t)) - F(0_i, 1_k, \boldsymbol{X}(t))\right]$.

Then $I_k(t)_{X_i(t)=0}$ is the effect of component k on the system lifetime of the failed component i.

When component i is restored from state 0 to state 1, due to which the change in system lifetime is $F(0_i, \boldsymbol{X}(t)) - F(1_i, \boldsymbol{X}(t))$, we can obtain

$$\begin{aligned}\frac{d\left[F(0_i, \boldsymbol{X}(t)) - F(1_i, \boldsymbol{X}(t))\right]}{dt} =& \lambda_k(t)R_k(t)\left[F(0_i, 0_k, \boldsymbol{X}(t)) - F(0_i, 1_k, \boldsymbol{X}(t))\right] \\ &- \lambda_k(t)R_k(t)\left[F(1_i, 0_k, \boldsymbol{X}(t)) - F(1_i, 1_k, \boldsymbol{X}(t))\right] \\ =& \lambda_k(t)R_k(t)\left[F(0_i, 0_k, \boldsymbol{X}(t)) + F(1_i, 1_k, \boldsymbol{X}(t))\right. \\ &\left. - F(0_i, 1_k, \boldsymbol{X}(t)) - F(1_i, 0_k, \boldsymbol{X}(t))\right] \\ =& \lambda_k(t)R_k(t)I_{i,k}^f. \end{aligned} \tag{4.48}$$

Let $I_k(t)_{X_i(t)} = I_k(t)_{X_i(t)=0} - I_k(t)_{X_i(t)=1}$. Then $I_k(t)_{X_i(t)}$ is the effect of component k on the system lifetime when component i is restored from state 0 to state 1.

In practice, improperly increasing the reliability of a system may increase losses. Therefore, to address the issue of minimising system failure losses during the life cycle of a production system, it is necessary to consider the costs in conjunction with the failures of different components.

Let $I_{i,k_{X_i(t)=0}}^f = c_{k|X_i(t)=0}\left(F(0_i, 0_k, \boldsymbol{X}(t)) - F(0_i, 1_k, \boldsymbol{X}(t))\right)$ represent the losses due to failures per unit of time caused by component k when component i is failed. The $c_{k|X_i(t)=0}$ is the maintenance cost of improving component k when component i is at state 0.

Then, let $I_{i,k_{X_i(t)=1}}^f = c_{k|X_i(t)=1}\left(F(1_i, 0_k, \boldsymbol{X}(t)) - F(1_i, 1_k, \boldsymbol{X}(t))\right)$ represent the losses from failures per unit of time caused by component k when component i is working. The $c_{k|X_i(t)=1}$ is the maintenance cost of improving component k when component i is at state 1.

Thus the importance of the joint loss can be defined as

$$I_1(t)_{X_i(t)} = I_{i,k_{X_i(t)=0}}^f - I_{i,k_{X_i(t)=1}}^f, \tag{4.49}$$

and

$$\begin{aligned} I_1(i,k)_{X_i(t)} =& c_{k|X_i(t)=1}\left(\left(F(0_i, \mathbf{0}_k, \boldsymbol{X}(t)) - F(0_i, \mathbf{1}_k, \boldsymbol{X}(t))\right)\right) \\ &- c_{k|X_i(t)=0}\left(\left(F(1_i, \mathbf{0}_k, \boldsymbol{X}(t)) - F(1_i, \mathbf{1}_k, \boldsymbol{X}(t))\right)\right). \end{aligned} \tag{4.50}$$

Equation (4.50) describes the contribution of component k to the change in system losses when component i is repaired or PM is performed on the component.

Example 4.3 If component i is a critical component, then when it is at state 0, the system will fail after component i fails. Without considering the repair time, if component k is repaired or preventively maintained, when the system has failed, regardless of whether the component k is a critical component, it will not incur system cost. As a result, we only need to consider maintenance cost. That is, if component k is a non-critical component, only maintenance cost is incurred. Therefore, when a component fails, we have

$$c_{k|X_i(t)=0} = c_k \Pr\{\phi(0_i, \mathbf{1}_i) < K\} + \Pr\{\phi(0_i, \mathbf{1}_i) \geq K\} \left[c_{s,k} \Pr\left\{\phi\left(0_i, 0_k, \mathbf{1}_{i,k}\right) < K\right\} + c_k\right], \tag{4.51}$$

where c_k is the cost incurred due to maintaining component k and $c_{s,k}$ is the system cost of maintaining component k. $(0_i, \mathbf{1}_i)$ represents that component i stops working and all the other components are working, and $\phi(0_i, \mathbf{1}_i) < K$ means that when component i fails, the state of the system will be lower than the failure threshold K, i.e., the system fails. $\Pr\{\phi(0_i, \mathbf{1}_i) < K\}$ ensures that the system will not fail when the component i is selected for PM. In addition to being a critical component, if component i is a non-critical component, then $\Pr\{\phi(0_i, \mathbf{1}_i) < K\} + \Pr\{\phi(0_i, \mathbf{1}_i) \geq K\} = 1$.

In addition, if component i does not fail, then it is only necessary to consider whether the system fails during the maintenance of component k, which results in system costs. In this case, the key is whether component k is a critical component. Therefore, we have

$$c_{k|X_i(t)=1} = c_{s,k} \Pr\{\phi(0_i, \mathbf{1}_i) < K\} + c_k. \tag{4.52}$$

Example 4.4 The costs of PM and CM for component k are different in practical application situations. We can distinguish whether component k is undergoing PM or CM based on the reliability of component k. Define c_k^{pf} as the cost of PM for component k, and c_k^f as the maintenance cost for component k failure. Then Eqs. (4.51) and (4.52) can be replaced by

$$c_{k|X_i(t)=0} = \left(\theta_k c_k^f + (1-\theta_k)\, c_k^{pf}\right) \Pr\{\phi(0_i, \mathbf{1}_i) < K\} + \Pr\{\phi(0_i, \mathbf{1}_i) \geq K\} \left[c_{s,k} \Pr\left\{\phi\left(0_i, 0_k, \mathbf{1}_{i,k}\right) < K\right\} + \theta_k c_k^f + (1-\theta_k)\, c_k^{pf}\right], \tag{4.53}$$

and

$$c_{k|X_i(t)=1} = c_{s,k} \Pr\left\{\phi\left(0_i, 0_k, \mathbf{1}_{i,k}\right) < K\right\} + \theta_k c_k^f + (1-\theta_k)\, c_k^{pf}, \tag{4.54}$$

where θ_k is a 0-1 variable that can be used to determine if a component k is a failure based on its reliability threshold.

Example 4.5 The relationship between reliability and cost for each component can be obtained based on historical data for similar components. However, in many cases such data is not available. Specifically, cost is a monotonically decreasing function of component reliability. The cost distribution function can be given as

$$c_k(t) = a_k \exp\left\{-\left[(1-f_k)\frac{R_{k(t)} - R_{k,\min}}{R_{k,\max} - R_{k(t)}}\right]\right\}, \tag{4.55}$$

where f_k is the feasibility of improving the reliability of component k, which is assumed to have a value between 0 and 1. $R_{k,\min}$ represents the minimum reliability of component k, $R_{k,\max}$ represents the maximum reliability of component k, and $a_k$is the corresponding cost factor for each component.

The feasibility parameter f_k is a constant that indicates the difficulty of improving the reliability of a component related to other components in the system. Many authors have proposed weighting factors for assigning reliability to quantify feasibility. These weights depend on certain influencing factors. Other literature also summarises some complex approaches such as the complexity of components, the level of technology, operational status, criticality, etc.

The cost of maintenance according to the cost allocation function can be expressed as

$$\begin{aligned} c_k(t) =& a_k \exp\left\{-[(1-f_k)\frac{R_k^{am}(t) - R_{k,\min}}{R_{k,\max} - R_k^{am}(t)}]\right\} \\ &- a_k \exp\left\{-[(1-f_k)\frac{R_k^{bm}(t) - R_{k,\min}}{R_{k,\max} - R_k^{bm}(t)}]\right\}, \end{aligned} \tag{4.56}$$

where $R_k^{am}(t)$ is the reliability of component k after repair and $R_k^{bm}(t)$ indicates the reliability before repair.

According to the above definition, the occurrence of the failure of component i will affect the feasibility factor in the maintenance cost function of component k. To better illustrate the effect of the above combination of cost functions, we give the following analysis

$$\begin{aligned} c_k(t)_{X_i(t)=0} =& a_k \exp\left\{-\left[\left(1 - f_{k|X_i(t)=0}\right)\frac{R_k^{am}(t) - R_{k,\min}}{R_{k,\max} - R_k^{am}(t)}\right]\right\} \\ &- a_k \exp\left\{-\left[\left(1 - f_{k|X_i(t)=0}\right)\frac{R_k^{bm}(t) - R_{k,\min}}{R_{k,\max} - R_k^{bm}(t)}\right]\right\}, \end{aligned} \tag{4.57}$$

and

$$\begin{aligned} c_k(t)_{X_i(t)=1} =& a_k \exp\left\{-\left[\left(1 - f_{k|X_i(t)=0}\right)\frac{R_k^{am}(t) - R_{k,\min}}{R_{k,\max} - R_k^{am}(t)}\right]\right\} \\ &- a_k \exp\left\{-\left[\left(1 - f_{k|X_i(t)=0}\right)\frac{R_k^{bm}(t) - R_{k,\min}}{R_{k,\max} - R_k^{bm}(t)}\right]\right\}, \end{aligned} \tag{4.58}$$

where $f_{k|X_i(t)} = 0$ and $f_{k|X_i(t)} = 1$ are the feasibility of repairing component k when it fails and when it works, respectively.

4.2.2 Component PM on the Expected Losses

According to Sect. 4.2.1, considering the second-order joint effect, we can obtained

$$I_2(i,k)_{X_i(t)=0} = \lambda_k(t) R_k(t)_{X_i(t)=0} I^f_{i,k_{X_i(t)=0}}. \tag{4.59}$$

$I_2(i,k)_{X_i(t)=0}$ can be used to estimate the contribution of component k to prolonging the remaining useful life of the system because it avoids maintenance costs and system losses without failures. Similarly, when the component is in state 1, the contribution of maintaining component k to the system can be given by

$$I_2(i,k)_{X_i(t)=1} = \lambda_k(t) R_k(t)_{X_i(t)=1} I^f_{i,k_{X_i(t)=1}}. \tag{4.60}$$

To better reflect the combined impact of the reliability of the two components on the system, it is possible to assess the reliability of the other component when one component is in a working or failed state. Jafary et al. [13] proposed a measure for the correlation between the reliability impact of two components.

$$\begin{aligned} R_k(t)_{X_i(t)=1} =& E\left[R_k(t)|R_i(t)=1\right] \\ =& \frac{p_i(t)p_k(t)+\rho_{i,k}\sigma_i(t)\sigma_k(t)}{p_i(t)}, \end{aligned} \tag{4.61}$$

where $\sigma_i(t)=\sqrt{Var\left[R_i(t)\right]}=\sqrt{p_i(t)q_i(t)}$, $p_i(t)+q_i(t)=1$. $R_i(t)=1$ if component i is working and $R_i(t)=0$ otherwise. For simplicity, success and failure are alternatively denoted with $p_i(t)$ and $q_i(t)$. The correlation between a pair of components i and k is denoted by $\rho_{i,k}$.

Then multiplying $\sigma_i(t)$ by the numerator and denominator in Eq. (4.61), we obtain

$$\begin{aligned} R_k(t)_{X_i(t)=1} =& p_k(t) + \frac{\rho_{i,k}\sigma_i^2(t)\sigma_k(t)}{\sigma_i(t)p_i(t)} \\ =& p_k(t) + \frac{\rho_{i,k}\sigma_k(t)p_i(t)q_i(t)}{\sigma_i(t)p_i(t)} \\ =& p_k(t) + \frac{\rho_{i,k}\sigma_k(t)q_i(t)}{\sigma_i(t)}. \end{aligned} \tag{4.62}$$

Similarly, we have

$$\begin{aligned} R_k(t)_{X_i(t)=0} =& E\left[R_k(t)|R_i(t)=0\right] \\ =& \frac{p_i(t)p_k(t)+\rho_{i,k}\sigma_i(t)\sigma_k(t)}{q_i(t)} \\ =& \frac{(1-q_i(t))p_k(t)+\rho_{i,k}\sigma_i(t)\sigma_k(t)}{q_i(t)} \end{aligned}$$

$$
\begin{aligned}
&= \frac{p_k(t)}{q_i(t)} - p_k(t) + \frac{+\rho_{i,k}\sigma_i(t)\sigma_k(t)}{q_i(t)} \\
&= \frac{p_k(t)}{q_i(t)} - p_k(t) + \frac{\rho_{i,k}\sigma_i^2(t)\sigma_k(t)}{\sigma_i(t)q_i(t)} \\
&= \frac{p_k(t)}{q_i(t)} - p_k(t) + \frac{\rho_{i,k}\sigma_k(t)p_i(t)q_i(t)}{\sigma_i(t)q_i(t)} \\
&= \frac{p_k(t)}{q_i(t)} - p_k(t) + \frac{p_i(t)\rho_{i,k}\sigma_k(t)}{\sigma_i(t)}.
\end{aligned}
\tag{4.63}
$$

When it is difficult to observe the real state of component i, Eq. (4.63) is used to obtain the state information of component i through component k. The states of component k and i are associated, and the state of component k is easily observed.

Therefore, the importance of the joint lifetime can be defined as

$$
\begin{aligned}
I_2(i,k)_{X_i(t)} =& I_2(i,k)_{X_i(t)=0} - I_2(i,k)_{X_i(t)=1} \\
=& \lambda_k(t)R_k(t)_{X_i(t)=0}I^f_{i,k_{X_i(t)=0}} - \lambda_k(t)R_k(t)_{X_i(t)=1}I^f_{i,k_{X_i(t)=1}} \\
=& \lambda_k(t)R_k(t)_{X_i(t)=0}c_{k|X_i(t)=0}\left((F(0_i,\mathbf{0}_k,\boldsymbol{X}(t)) - F(0_i,\mathbf{1}_k,\boldsymbol{X}(t))\right) \\
=& \lambda_k(t)R_k(t)_{X_i(t)=1}c_{k|X_i(t)=1}\left((F(1_i,\mathbf{0}_k,\boldsymbol{X}(t)) - F(1_i,\mathbf{1}_k,\boldsymbol{X}(t))\right).
\end{aligned}
\tag{4.64}
$$

$I_2(i,k)_{X_i(t)}$ describes the contribution of component k to the system when repairing or performing PM on component i. The contribution of maintaining the component to maximise the life of the system and avoiding loss of failure can be expressed as $I_3(i)_{X_i(t)}$. Then we can obtain the expected lifetime importance.

$$
\begin{aligned}
I_3(i)_{X_i(t)} =& \sum_{k=1,k\neq i}^{n} \{\lambda_k(t)\left[R_k(t)_{X_i(t)=0}I^f_{i,k(X_i(t)=0)} - R_k(t)_{X_i(t)=1}I^f_{i,k(X_i(t)=1)}\right]\} \\
&\times \sum_{k=1,k\neq i}^{n} \left[\begin{array}{l} \lambda_k(t)R_k(t)_{X_i(t)=0}c_{k|X_i(t)=0}\left((F(0_i,\mathbf{0}_k,\boldsymbol{X}(t)) - F(0_i,\mathbf{1}_k,\boldsymbol{X}(t))\right) \\ -\lambda_k(t)R_k(t)_{X_i(t)=1}c_{k|X_i(t)=1}\left((F(1_i,\mathbf{0}_k,\boldsymbol{X}(t)) - F(1_i,\mathbf{1}_k,\boldsymbol{X}(t))\right) \end{array}\right] \\
=& \sum_{k=1,k\neq i}^{n} I_2(i,k)_{X_i(t)}
\end{aligned}
\tag{4.65}
$$

It represents the expected contribution of maintenance component i to extend the system life. Based on the three different cost scenarios, we need to further analyse.

Scenario 1 Based on the positive and negative correlation reliability effects between the two components and Eqs. (4.53) and (4.54), we can derive Eqs. (4.66) and (4.67) as

$$c_{k|X_i(t)=0} = c_k \Pr\left\{R_U\left(\frac{p_1(t)}{q_i(t)} - p_1(t) + \frac{p_i(t)\rho_{i,1}\sigma_1(t)}{\sigma_i(t)}, \frac{p_2(t)}{q_i(t)} - p_2(t)\right.\right.$$
$$\left.\left. + \frac{p_i(t)\rho_{i,2}\sigma_2(t)}{\sigma_i(t)}, \cdots, \frac{p_j(t)}{q_i(t)} - p_j(t) + \frac{p_i(t)\rho_{i,j}\sigma_j(t)}{\sigma_i(t)}\right) < \tau\right\}$$
$$+ \Pr\left\{R_U\left(\frac{p_1(t)}{q_i(t)} - p_1(t) + \frac{p_i(t)\rho_{i,1}\sigma_1(t)}{\sigma_i(t)}, \frac{p_2(t)}{q_i(t)} - p_2(t) + \frac{p_i(t)\rho_{i,2}\sigma_2(t)}{\sigma_i(t)},\right.\right.$$
$$\left.\left. \cdots, \frac{p_j(t)}{q_i(t)} - p_j(t) + \frac{p_i(t)\rho_{i,j}\sigma_j(t)}{\sigma_i(t)}\right) \geq \tau\right\}\left[c_{s,k} \Pr\left\{R_U\left(\frac{p_1(t)}{q_k(t)} - p_k(t)\right.\right.\right.$$
$$+ \frac{p_k(t)\rho_{k,1}\sigma_1(t)}{\sigma_k(t)}, \frac{p_2(t)}{q_k(t)} - p_2(t) + \frac{p_k(t)\rho_{k,2}\sigma_2(t)}{\sigma_k(t)}, \ldots, \frac{p_j(t)}{q_k(t)} - p_j(t),$$
$$\left.\left.\left. + \frac{p_k(t)\rho_{k,j}\sigma_j(t)}{\sigma_k(t)}\right) < \tau\right\} + c_s\right] \quad j \in \{1, 2, \ldots, n\}, j \neq i \ \&\ j \neq k, \tag{4.66}$$

and

$$c_{k|X_i(t)=1} = \left[c_{s,k} \Pr\left\{R_U\left(\frac{p_1(t)}{q_k(t)} - p_k(t) + \frac{p_k(t)\rho_{k,1}\sigma_1(t)}{\sigma_k(t)}, \frac{p_2(t)}{q_k(t)} - p_2(t)\right.\right.\right.$$
$$\left.\left.\left. + \frac{p_k(t)\rho_{k,2}\sigma_2(t)}{\sigma_k(t)}, \ldots, \frac{p_j(t)}{q_k(t)} - p_j(t) + \frac{p_k(t)\rho_{k,j}\sigma_j(t)}{\sigma_k(t)}\right) < \tau\right\} + c_k\right],$$
$$j \in \{1, 2, \ldots, n\}, j \neq k, \tag{4.67}$$

where $R_U(t)$ represents the reliability of the system at time t, and τ is the system reliability threshold. When the reliability of the system is lower than the threshold τ, the system fails. We can calculate the reliability of component k by observing the reliability of component i. Similarly, we can calculate the reliability of other components related to component i, based on which one can obtain the reliability of the system at time t.

Scenario 2 We can derive Eqs. (4.68) and (4.69) as

$$c_{k|X_i(t)=0} = c_k \Pr\left\{R_U\left(\frac{p_1(t)}{q_i(t)} - p_1(t) + \frac{p_i(t)\rho_{i,1}\sigma_1(t)}{\sigma_i(t)}, \frac{p_2(t)}{q_i(t)} - p_2(t)\right.\right.$$
$$\left.\left. + \frac{p_i(t)\rho_{i,2}\sigma_2(t)}{\sigma_i(t)}, \cdots, \frac{p_j(t)}{q_i(t)} - p_j(t) + \frac{p_i(t)\rho_{i,j}\sigma_j(t)}{\sigma_i(t)}\right) < \tau\right\}$$
$$+ \Pr\left\{R_U\left(\frac{p_1(t)}{q_i(t)} - p_1(t) + \frac{p_i(t)\rho_{i,1}\sigma_1(t)}{\sigma_i(t)}, \frac{p_2(t)}{q_i(t)} - p_2(t) + \frac{p_i(t)\rho_{i,2}\sigma_2(t)}{\sigma_i(t)},\right.\right.$$
$$\left.\left. \cdots, \frac{p_j(t)}{q_i(t)} - p_j(t) + \frac{p_i(t)\rho_{i,j}\sigma_j(t)}{\sigma_i(t)}\right) \geq \tau\right\}\left[c_{s,k} \Pr\left\{R_U\left(t|\frac{p_1(t)}{q_k(t)} - p_k(t)\right.\right.\right.$$
$$+ \frac{p_k(t)\rho_{k,1}\sigma_1(t)}{\sigma_k(t)}, \frac{p_2(t)}{q_k(t)} - p_2(t) + \frac{p_k(t)\rho_{k,2}\sigma_2(t)}{\sigma_k(t)}, \ldots, \frac{p_j(t)}{q_k(t)}$$

$$- \; p_j(t) + \frac{p_k(t)\rho_{k,j}\sigma_j(t)}{\sigma_k(t)}\Bigg) < \tau \Bigg\} + \theta_k c_k^f + (1-\theta_k)c_k^{pf}\Bigg],$$
$$j \in \{1,2,\ldots,n\}, j \neq i \,\&\, j \neq k, \tag{4.68}$$

and

$$c_{k|X_i(t)=1} = \Bigg[c_{s,k} \Pr\Bigg\{R_U\Bigg(p_1(t) + \frac{q_k(t)\rho_{k,1}\sigma_1(t)}{\sigma_k(t)}, p_2(t) + \frac{q_k(t)\rho_{k,2}\sigma_2(t)}{\sigma_k(t)}, \ldots, p_j(t) + \frac{q_k(t)\rho_{k,j}\sigma_j(t)}{\sigma_k(t)}\Bigg) < \tau\Bigg\} + \theta_k c_k^f + (1-\theta_k)c_k^{pf}\Bigg], \tag{4.69}$$

where $j \in \{1, 2, \ldots, n\}$ and θ_k is a 0-1 variable, it can be used to choose whether to perform PM on the state of component k.

Scenario 3 $R_{i,\min}$ is the minimum reliability of component i. $R_{i,\max}$ is the maximum achievable reliability of the component i. For Eqs. (4.62) and (4.63), we have $R_{i,\min} = \min(R_k(t)_{X_i(t)=0})$ and $R_{i,\max} = \max(R_k(t)_{X_i(t)=0})$ when component i failed. Similarly, $R_{i,\max} = \max(R_k(t)_{X_i(t)=1})$ when component i is reliable. Denote $Q_i^{am}(t) = \frac{p_k^{am}(t)}{q_i(t)} - p_k^{am}(t) + \frac{p_i(t)\rho_{i,k}\sigma_k(t)}{\sigma_i(t)}$ and $Q_i^{bm}(t) = \frac{p_k^{bm}(t)}{q_i(t)} - p_k^{bm}(t) + \frac{p_i(t)\rho_{i,k}\sigma_k(t)}{\sigma_i(t)}$.
Then we have

$$\begin{aligned}
c_{k|X_i(t)=0} =& a_k \exp\left\{\left[\left(1 - f_{k|X_i(t)=0}\right) + \frac{Q_i^{am}(t) - \min\left(R_k(t)_{X_i(t)=0}\right)}{\max\left(R_k(t)_{X_i(t)=0}\right) - \left(Q_i^{am}(t)\right)}\right]\right\} \\
& - a_k \exp\left\{\left[\left(1 - f_{k|X_i(t)=0}\right) + \frac{Q_i^{bm}(t) - \min(R_k(t)_{X_i(t)=0})}{\max\left(R_k(t)_{X_i(t)=0}\right) - (Q_i^{bm}(t))}\right]\right\} \\
& \times \max\left(R_k(t)_{X_i(t)=0}\right) - (Q_i^{bm}(t)) \\
& \times \frac{Q_i^{bm}(t) - \min(R_k(t)_{X_i(t)=0})}{\max\left(R_k(t)_{X_i(t)=0}\right) - (Q_i^{bm}(t))} \\
& \times c_{k|X_i(t)=1} = a_k e^{\left[\left(1 - f_{k|X_i(t)=1}\right) + \frac{Q_i^{am}(t) - \min\left(R_k(t)_{X_i(t)=0}\right)}{\max\left(R_k(t)_{X_i(t)=0}\right) - \left(Q_i^{am}(t)\right)}\right]} \\
& - a_k e^{\left[\left(1 - f_{k|X_i(t)=1}\right) + \frac{Q_i^{bm}(t) - \min(R_k(t)_{X_i(t)=0})}{\max\left(R_k(t)_{X_i(t)=0}\right) - (Q_i^{bm}(t))}\right]}
\end{aligned} \tag{4.70}$$

While component i is being repaired, given a fixed repair cost C and fixed maintenance time t, we should determine the components for PM to minimise losses. On the other hand, we need to consider the impact of the change in the number of repairmen on the PM decision. According to the classification of the different cost scenarios, integer programming should also be discussed in these three scenarios.

Scenario 1:

$$\max \sum_{k=1,k\neq i}^{n} I_2(i,k)_{X_i(t)} z_k, \tag{4.71}$$

$$s.t. \begin{cases} z_k \in \{0,1\} \\ c_i + \sum_{k=1,k\neq i}^{n} \begin{bmatrix} c_k \Pr\{\phi(0_i, \mathbf{1}_i) < K\} + \Pr\{\phi(0_i, \mathbf{1}_i) \geq K\} \\ \left[c_{s,k} \Pr\left\{\phi\left(0_i, 0_k, \mathbf{1}_{i,k}\right) < K\right\} + c_k\right] \end{bmatrix} z_k \leq C \\ \sum_{k=1,k\neq i}^{n} \left(t_{k0}^{pf} + \varepsilon t_k^{pf}\right) z_k \leq t_{i0}^{f} + \delta t_i^{f}, \varepsilon \in [0,1], \delta \in [0,1] \end{cases}.$$

Scenario 2:

$$\max \sum_{k=1,k\neq i}^{n} I_2(i,k)_{X_i(t)} z_k, \tag{4.72}$$

s.t.

$$\begin{cases} z_k \in \{0,1\} \\ c_i + \sum_{k=1,k\neq i}^{n} \begin{bmatrix} c_k \Pr\{\phi(0_i, \mathbf{1}_i) < K\} + \Pr\{\phi(0_i, \mathbf{1}_i) \geq K\} \\ \left[c_{s,k} \Pr\left\{\phi\left(0_i, 0_k, \mathbf{1}_{i,k}\right) < K\right\} + \theta_k c_k^{f} + (1-\theta_k) c_k^{pf}\right] \end{bmatrix} z_k \leq C \\ \sum_{k=1,k\neq i}^{n} \left(t_{k0}^{pf} + \varepsilon t_k^{pf}\right) z_k \leq t_{i0}^{f} + \delta t_i^{f}, \varepsilon \in [0,1], \delta \in [0,1] \end{cases}.$$

Scenario 3:

$$\max \sum_{k=1,k\neq i}^{n} I_2(i,k)_{X_i(t)} z_k, \tag{4.73}$$

s.t.

$$\begin{cases} z_k \in \{0,1\} \\ c_i + \sum_{k=1,k\neq i}^{n} \left[a_k e^{(1-f_k)\frac{R_k^{am}(t) - R_{k,\min}}{R_{k,\max} - R_k^{am}(t)}} - a_k e^{(1-f_k)\frac{R_k^{bm}(t) - R_{k,\min}}{R_{k,\max} - R_k^{bm}(t)}} \right] z_k \leq C \\ \sum_{k=1,k\neq i}^{n} \left(t_{k0}^{pf} + \varepsilon t_k^{pf}\right) z_k \leq t_{i0}^{f} + \delta t_i^{f}, \varepsilon \in [0,1], \delta \in [0,1] \end{cases},$$

where c_i represents the maintenance cost for component i, z_k is the decision variable represents whether component k should be maintained or not, T represents the total maintenance time, t_i^f and t_k^{pf} represent the maintenance time of component i and the PM time of component k, respectively, and t_{i0}^f and t_{k0}^{pf} represent the necessary time required for CM and PM of the corresponding component, respectively. δ and ε are the influence factors of CM personnel and PM personnel, respectively.

In the three different cost scenarios that are discussed above, different cost constraint functions reflect different $c_{k|X_i(t)}$. It is necessary to maximise the contribution of the remaining components to system when component i fails. It also satisfies that all repairable maintenance costs and PM costs are less than the total maintenance cost. Meanwhile the total PM time is less than the CM time.

Note that z_k can only take values from 0 or 1. After a failed component being removed, there are $n - 1$ components left. Therefore, it suffices to test the 2^{n-1} combinations of z_k for deriving the optimal maintenance policy. Under the optimal maintenance policy $\{z_k^*, k \neq i\}$, the number of maintained components is $\sum_{k \neq i} z_k^*$.

4.3 Component Importance Measures for Systems with Different Maintenance Policies

In the life cycle of a system, degradation of the system is inevitable due to the environmental interference and the wear and tear of its own operation. When the performance degradation reaches a certain level, the system may not function properly. At this point, maintenance is needed to restore the system. Maintenance should vary from component to component throughout the life cycle due to the performance differences between components. At the same time, due to the interaction between components, maintenance sequence of different components may have significant differences in improving system reliability. Therefore, in order to improve the reliability of the system and reduce the maintenance cost of the cycle, it is crucial to develop certain measures to quantify the importance of components and guide the selection of components for PM, so as to improve the reliability of the system.

Various methods are proposed to guide the maintenance for complex systems. With Bayesian networks, Cai et al. [5] combined the Bayesian network diagnosis stage with the reliability evaluation stage and proposed a reliability evaluation method to study the value of reliability of the system and its changing trend when components fail. Sun et al. [19] proposed a complex electronic system evaluation method combining the failure physical method with a copula Bayesian network to guide the improvement of system reliability and system maintenance. Considering the resilience of the system, Xing [25] proposed a reliability analysis method for cascading fault modelling and constructed mitigation strategies for the resilience system to deal with cascading failures. Xu et al. [26] proposed a resilience-based multi-state network component importance ranking method to determine the importance of each component in the network. In terms of optimisation policy, Levitin et al. [15] proposed a task success probability evaluation method based on event transfer, which solved the problem of optimal preventive replacement scheduling. Consideration the maintenance cost of the system and its components, Wu and Coolen [21] proposed a cost-based importance measure method that considers the cost of maintaining the system and its components in a limited time. Dui et al. [7] proposed a cost-based integrated importance measure (IIM) to identify components or component groups for PM, considering the combined impact of maintenance cost and maintenance time on system reliability. Xiao et al. [24] proposed an effective simulation budget allocation policy to determine how to balance load and protect system components to maximise the probability of correct sequencing. Feng et al. [8] proposed a comprehensive evaluation method based on Monte Carlo to evaluate the reliability and performance of heterogeneous components in complex human-computer systems.

4.3.1 The Age Replacement Maintenance Policy Based on Importance of Maintenance Cost

If a component fails before a pre-specified time T, the component is subsequently replaced and this kind of repair is called post-maintenance. Under the condition of the age replacement maintenance policy, a system is always repaired when it reaches its age threshold T or fails.

A maintenance cycle is defined as the time required for a component to reach its PM or subsequent maintenance after PM or post-repair. From this, the expected operational time of the component in a maintenance cycle is:

$$\begin{aligned} E(U) &= \int_0^T xf(x)dx + TR(T) \\ &= \int_0^T R(x)dx. \end{aligned} \tag{4.74}$$

A component is repaired in two ways. One is the way of repair upon failures, and the probability is $F(t)$. The other is PM, and the probability is $R(t) = 1 - F(t)$. So, the expected replacement time of the component in a maintenance cycle is:

$$E(\tilde{V}) = \tau_f F(T) + \tau_p R(T). \tag{4.75}$$

There are also two corresponding maintenance cycles. The probability that performing PM after time T is reached is $R(t)$, and the probability that performing post-maintenance before time T is reached is $F(t)$. Therefore, the expected cycle length is the summation of the expected working time of the system and the expected repair time of the component.

$$\begin{aligned} P^{ECL}(T) &= E(U) + E(V) \\ &= \int_0^T R(x)dx + \tau_f F(T) + \tau_p R(T). \end{aligned} \tag{4.76}$$

The expected cycle cost can also be calculated in a similar way:

$$P^{ECC}(T) = c_f F(T) + c_p R(T). \tag{4.77}$$

Because the integrated importance needs to consider the probability of being in each state, the transition rate is the inverse of the cycle, and its expression is:

$$\lambda_i(T) = \frac{1}{P^{ECL}(T)} = \frac{1}{\int_0^T R(t)dt + \tau_f F(T) + \tau_p R(T)}. \tag{4.78}$$

The probability that the system stays in the working state is the expected ratio of the expected working time to the length of the cycle:

$$R_i(T) = \frac{E(U)}{P^{ECL}(T)} = \frac{\int_0^T R(x)dx}{\int_0^T R(x)dx + \tau F(T) + \tau R(T)}. \tag{4.79}$$

The integrated importance is given by

$$\begin{aligned} I_i^{im}(T) &= \lambda_i(T) R_i(T) I_i^{\mathrm{B}}(T) \\ &= \frac{1}{P^{\mathrm{ECL}}(T)} \cdot \frac{E(U)}{P^{\mathrm{ECL}}(T)} I_i^{\mathrm{B}}(T) \\ &= \frac{E(U) I_i^{\mathrm{B}}(T)}{\left(P^{\mathrm{ECL}}(T)\right)^2}. \end{aligned} \tag{4.80}$$

The maintenance cost per unit of time of component i in the steady state of the system is:

$$J_\infty(T) = \frac{P^{ECC}(T)}{P^{ECL}(T)} = \frac{c_f F(T) + c_p R(T)}{\int_0^T R(t)dt + \tau_f F(T) + \tau_p R(T)}. \tag{4.81}$$

The integrated importance of available components based on maintenance cost is:

$$\begin{aligned} I_i^{cim}(T) &= \frac{\partial C(T)}{\partial I_i^{\mathrm{im}}(T)} \\ &= \frac{t J_\infty(T)}{I_i^{\mathrm{im}}(T)} \\ &= t \frac{P^{ECC}(T)}{P^{ECL}(T)} \frac{\left(P^{\mathrm{ECL}}(T)\right)^2}{E(U) I_i^{\mathrm{B}}(T)} \\ &= t \frac{P^{ECC}(T) P^{ECL}(T)}{E(U) I_i^{\mathrm{B}}(T)}, \end{aligned} \tag{4.82}$$

where $P^{ECC}(T) = c_f F(T) + c_p R(T)$, $P^{ECL}(T) = \int_0^T R(x)dx + \tau_f F(T) + \tau_p R(T)$, and $E(U) = \int_0^T R(x)dx$.

Proposition 4.1 *In a series system, if the components are independent of each other and obey the exponential distribution, the integrated importance of component i considered maintenance cost is:*

$$I_i^{cim}(t) = \frac{E(U)\exp(-t\sum_{k=1,k\neq i}^{n}\lambda_k)}{\left(P^{ECL}(T)\right)^2}. \tag{4.83}$$

Proof In a series system, any component failure will cause the system to fail. Assuming that it is impossible for more than one component to fail in a very short time and that only the system failure caused by a single component is considered, the maintenance cost of component i in time period $(0, t)$ is:

$$C^i(t) = tJ_\infty(T) = t\frac{P^{ECC}(T)}{P^{ECL}(T)}. \tag{4.84}$$

In a series system, the Birnbaum importance of component i is:

$$I_i^{\mathrm{B}}(t) = \exp(-t\sum_{k=1,k\neq i}^{n}\lambda_k). \tag{4.85}$$

Therefore, for any component i, its integrated importance based on maintenance cost is:

$$\begin{aligned} I_i^{cim}(t) &= \frac{\partial C(t)}{\partial I_i^{\mathrm{im}}(t)} \\ &= \frac{t\frac{P^{ECC}(T)}{P^{ECL}(T)}}{t\frac{P^{ECC}(T)P^{ECL}(T)}{E(U)I_i^{\mathrm{B}}(T)}} \\ &= \frac{E(U)\exp(-t\sum_{k=1,k\neq i}^{n}\lambda_k)}{\left(P^{\mathrm{ECL}}(T)\right)^2}, \end{aligned} \tag{4.86}$$

where

$$P^{ECL}(T) = \int_0^T R(x)dx + \tau_f F(T) + \tau_p R(T), \tag{4.87}$$

$$E(U) = \int_0^T R(x)dx, \tag{4.88}$$

Proposition 4.2 *In a parallel system, if the components are independent of each other and obey the exponential distribution, the integrated importance of component i based on the maintenance cost is:*

$$I_i^{cim}(t) = \frac{E(U)\sum_{k=1,k\neq i}^{n}(1-\exp(-\lambda_k t))}{\left(P^{ECL}(T)\right)^2}. \tag{4.89}$$

Proof In a parallel system, the system will fail if and only if all components fail. No more than one component fails at the same time in any short period of time, so the probability of system failure caused by component i is zero. Therefore, the maintenance cost of component i in the parallel system during time period $(0, t)$ is:

$$C^i(t) = tJ_\infty(T) = t\frac{P^{ECC}(T)}{P^{ECL}(T)}. \tag{4.90}$$

In a parallel system, the Birnbaum importance of a component is:

$$I_i^{\mathrm{B}}(t) = \sum_{k=1,k\neq i}^{n}(1-\exp(-\lambda_k t)). \tag{4.91}$$

For any component i, its integrated importance based on maintenance cost is:

$$\begin{aligned} I_i^{cim}(t) &= \frac{t\frac{P^{ECC}(T)}{P^{ECL}(T)}}{t\frac{P^{ECC}(T)P^{ECL}(T)}{E[U]I_i^{\mathrm{B}}(T)}} \\ &= \frac{E(U)}{\left(P^{\mathrm{ECL}}(T)\right)^2}\sum_{k=1,k\neq i}^{n}(1-\exp(-\lambda_k t)), \end{aligned} \tag{4.92}$$

where

$$E(U) = \int_0^T R(x)dx. \tag{4.93}$$

4.3.2 A PM Policy Based on Importance of Maintenance Cost

There are many factors that should be considered when formulating a PM policy. One factor can be the importance measure of the components. This is also an important decision variable for distinguishing different components of PM. Si et al. [18] proposed the integrated importance measure (IIM) to distinguish the priority of PM of different components. Therefore, we propose a cost-based PM policy based on IIM. It is defined as

$$I_i^{im}(t) = R_i(t)\lambda_i(t)\sum_{j=1}^{M}a_i\{\Pr[\Phi(1_i, \boldsymbol{X}(t)) = j] - \Pr[\Phi(0_i, \boldsymbol{X}(t)) = j]\}. \tag{4.94}$$

This is the IIM of component i, where $R_i(t)$ is the reliability of component i at time t, $\lambda_i(t)$ is the failure rate of component i at time t, and

$$\sum_{j=1}^{M} a_i\{\Pr[\Phi(1_i, \boldsymbol{X}(t)) = j] - \Pr[\Phi(0_i, \boldsymbol{X}(t)) = j]\} \tag{4.95}$$

is the contribution of component i performance degradation to the change of system performance. Therefore, Eq. (4.95) can also be written as

$$I_i^{im}(t) = R_i(t)\lambda_i(t)\frac{\partial R(t)}{\partial R_i(t)}. \tag{4.96}$$

The component failure rate can be expressed as $\lambda_i(t) = -\frac{dR_i(t)/dt}{R_i(t)}$. Therefore, the IIM of component i can be expressed as

$$I_i^{im}(t) = -\frac{dR_i(t)}{dt}\frac{\partial R(t)}{\partial R_i(t)}. \tag{4.97}$$

In fact, the sum of the IIM values of the n components measures the degradation of the reliability of the system per unit of time. The system reliability composed of n components can be expressed as $R(t) = f(R_1(t), R_2(t), \ldots, R_n(t))$.

$$\begin{aligned}\frac{dR(t)}{dt} &= \frac{df(R_1(t), R_2(t), \ldots, R_n(t))}{dt}\\ &= \sum_{i=1}^{n}\frac{dR_i(t)}{dt}\frac{\partial R(t)}{\partial R_i(t)}\\ &= -\sum_{i=1}^{n} I_i^{im}(t).\end{aligned} \tag{4.98}$$

According to Eq. (4.98), the reliability degradation of the system in a unit of time is the sum of the IIM values of the n components at time t. Therefore, the IIM value of component i is the contribution of component i to the degradation of the system reliability at time t. Therefore, the PM sequence of components can be determined by ordering the IIM values of components. A cost-based PM policy is therefore defined below.

Definition 4.6 A cost-based PM policy can be defined as:

$$I_i^{pm}(t) = \frac{C^{\mathrm{PM}}(t)}{I_i^{\mathrm{im}}(t)}. \tag{4.99}$$

The goal of the maintenance policy is to reduce the maintenance cost and improve the reliability of the system as much as possible. The IIM value of component i represents the contribution of component i to system reliability degradation. Therefore, the greater the IIM value of component i is, the greater its contribution to system reliability degradation will be. Therefore, it can be seen from Eq. (4.99) that the components with lower $I_i^{pm}(t)$ value should be preferred for PM, which can improve the system reliability with less maintenance cost.

The above is the cost-based PM policy. When the system reaches the PM time, components can be selected for PM based on this policy. This policy is mainly for system non-critical components. Next, we will introduce cost-based operation maintenance (OM) policy. This policy is aimed at maintenance decisions after the failure of key components of the system.

4.3.3 The Operation Maintenance Policy Based on Importance of Maintenance Cost

When a key component k in the system fails and needs to be shut down for maintenance, PM can be carried out at the same time after component k in the system shut down, so as to reduce the maintenance cost of the system life cycle. The maintenance policy at this time is an OM (operation maintenance) policy. An OM policy needs to determine the importance of the other components after the failure of component k, so a new binary importance measure is needed to determine the importance of the components.

On the basis of IIM, Dui et al. [6] proposed the joint integrated importance measure (JIIM) to determine the PM choice of the remaining components after the failure of the key component k in the system. This is a binary importance measure, and its definition is as follows

$$I_i^{jim}(t)_{X_k(t)} = I_i^{im}(t)_{X_k(t)=1} - I_i^{im}(t)_{X_k(t)=0}, \tag{4.100}$$

where $I_i^{im}(t)_{X_k(t)=1}$ represents the contribution of component i to system performance degradation when component k is in a perfect state. $I_i^{im}(t)_{X_k(t)=0}$ represents the contribution of component i to system performance degradation when component k is in a fault state. The difference $I_i^{jim}(t)_{X_k(t)}$ between them is defined as the JIIM of component i when component k is under maintenance. Equation (4.100) can be expanded and written as

$$
\begin{aligned}
& I_i^{jim}(t)_{X_k(t)} \\
& = R_i(t)\lambda_i(t)\sum_{j=1}^{M} a_j \left\{\Pr\left[\Phi(1_k, 1_i, \boldsymbol{X}(t)) = j\right] - \Pr\left[\Phi(1_k, 0_i, \boldsymbol{X}(t)) = j\right]\right\} \\
& \quad - R_i(t)\lambda_i(t)\sum_{j=1}^{M} a_j \left\{\Pr\left[\Phi(0_k, 1_i, \boldsymbol{X}(t)) = j\right] - \Pr\left[\Phi(0_k, 0_i, \boldsymbol{X}(t)) = j\right]\right\}.
\end{aligned}
\tag{4.101}
$$

The IIM describes the contribution of component i to system performance degradation, while the JIIM describes the contribution of component i to system performance degradation after the failure of component k, which is a binary importance measure. There is a corresponding relationship between them.

$$
\begin{aligned}
I_i^{jim}(t) &= R_i(t)\lambda_i(t)\sum_{j=1}^{M} a_j \left\{\Pr\left[\Phi(1_i, \boldsymbol{X}(t)) = j\right] - \Pr\left[\Phi(0_i, \boldsymbol{X}(t)) = j\right]\right\} \\
&= R_i(t)\lambda_i(t)\sum_{j=1}^{M} a_j \left\{\begin{array}{c} R_k(t)\Pr\left[\Phi(1_k, 1_i, \boldsymbol{X}(t)) = j\right] \\ +(1 - R_k(t))\Pr\left[\Phi(0_k, 1_i, \boldsymbol{X}(t)) = j\right] \\ -R_k(t)\Pr\left[\Phi(1_k, 0_i, \boldsymbol{X}(t)) = j\right] \\ -(1 - R_k(t))\Pr\left[\Phi(0_k, 0_i, \boldsymbol{X}(t)) = j\right] \end{array}\right\}.
\end{aligned}
\tag{4.102}
$$

Substitute Eqs. (4.101) and (4.102) into Eq. (4.94) to obtain

$$
I_i^{\text{im}}(t) = R_k(t) I_i^{\text{jim}}(t)_{X_k(t)} + I_i^{\text{im}}(t)_{X_k(t)=0}. \tag{4.103}
$$

When component k is critical, its failure causes the system to shut down for maintenance, then $I_i^{im}(t)_{X_k(t)=0} = 0$. So, the relationship between JIIM and IIM of component i has the following relationship

$$
I_i^{jim}(t)_{X_k(t)} = \frac{1}{R_k(t)} I_i^{\text{im}}(t). \tag{4.104}
$$

When the critical component k fails, Eq. (4.104) shows that the contribution of component i to the system reliability degradation can be expressed as

$$
I_i^{jim}(t)_{X_k(t)} = -\frac{dR_i(t)/dt}{R_k(t)} \frac{\partial R(t)}{\partial R_i(t)}. \tag{4.105}
$$

This is the JIIM definition of the impact of component i on system reliability degradation after component k fails. Next, a cost-based OM policy is given.

Definition 4.7 A cost-based OM policy can be defined as:

$$I_i^{om}(t) = \frac{C^{OM}(t)}{I_i^{jim}(t)_{X_k(t)}}, \tag{4.106}$$

where $C^{OM}(t)$ is the maintenance cost of OM, and $I_i^{jim}(t)_{X_k(t)}$ is the influence of component i on system reliability degradation after the key component k in the system fails. The larger the value of $I_i^{jim}(t)_{X_k(t)}$, the higher the improvement of system reliability. It can therefore guide the selection of components under OM. The OM policy is to use less maintenance costs to improve the system reliability as much as possible. Therefore, when the $I_i^{om}(t)$ value of a component is smaller, it indicates that the component should be maintenance first, which can reduce the maintenance cost of the whole life cycle of the system.

The above is the cost-based OM policy. When a system maintenance opportunity is obtained after the key component k fails, the OM policy can be used to select the remaining components for PM.

4.3.4 Cost-Based Risk Analysis

Cost is an important factor in the maintenance policy optimisation process. During regular system maintenance, improper maintenance intervals may cause system to be over- or under-maintained. Therefore, the cost-based risk analysis of the system is particularly important. While a system is being maintained, the expected loss for a given failure of a system with n components can be expressed as

$$\overline{C}_f(t) = \sum_{k=1}^{n} \frac{\lambda_k(t)}{\lambda_s(t)} \overline{C}_k(t), \tag{4.107}$$

where $\overline{C}_k$ is the expected loss when system component k fails, and $\lambda_s(t)$ is the system failure rate, which is related to the connections between components. Therefore, the risk of failure for a repairable system can be expressed as

$$\begin{aligned} K(t) &= (1 - R_s(t))\overline{C}_f(t) \\ &= (1 - R_s(t)) \sum_{k=1}^{n} \frac{\lambda_k}{\lambda_s} \overline{C}_k(t). \end{aligned} \tag{4.108}$$

By comparing the risk of failure K with the maintenance cost of the system, the relationship between the risk of failure and the maintenance cost in a certain period of time can be determined. When the risk of failure in a certain period of time is less than the maintenance cost of the system, it indicates that the risk of failure in this period of time is relatively small. In other words, the system is not prone to failure, so the maintenance frequency of components can be reduced during this period, thus reducing unnecessary maintenance costs of the system. When the risk of failure in a certain period of time is greater than the maintenance cost of the system, the risk of failure in this period of time is relatively large and the system is prone to failure at this stage, so the maintenance frequency of the system should be increased to reduce the probability of system failure. The system maintenance frequency is determined by cost-based risk analysis to ensure that the system maintenance is in the optimal state throughout the whole life cycle.

4.4 Summary

Considering the joint effect of component reliability and maintenance cost on the system reliability in a maintenance time, this chapter proposed a cost-based IIM to identify which components should be selected for preventive maintenance. For different maintenance policies, when some components fail, importance measures can be used to identify the important components that affect the system reliability. The proposed importance measures also take into account the component maintenance cost and maintenance cost associated with the system failure, providing more effective guidance in practical applications.

Maintenance policies considered in this chapter can be extended to other scenarios. Different maintenance costs, such as the constraint of other resources (finance, external factor), and opportunity cost, can also be taken into consideration.

References

1. Ahmadi R, Wu S (2018) An optimal maintenance policy based on partial information. In: Safety and reliability–safe societies in a changing world, pp 511–518. CRC Press
2. Annamraju S, Naikan V, Wu S (2019) Assessing the reliability of a repairable system in a usage or combined time scale. In: The 11th international conference on mathematical methods in reliability, Hong Kong
3. Bennett D, Jenney B (1980) Reliability: its implications in production systems design. Omega 8(4):433–440
4. Bohlooli-Zefreh M, Asadi M, Parvardeh A (2021) On the reliability and optimal maintenance of systems under a generalized mixed δ-shock model. Proc Inst Mech Eng, Part O: J Risk Reliab 235(5):909–922
5. Cai B, Liu Y, Fan Q, Zhang Y, Liu Z, Yu S, Ji R (2014) Multi-source information fusion based fault diagnosis of ground-source heat pump using Bayesian network. Appl Energy 114:1–9
6. Dui H, Si S, Zuo MJ, Sun S (2015) Semi-Markov process-based integrated importance measure for multi-state systems. IEEE Trans Reliab 64(2):754–765
7. Dui H, Si S, Yam RC (2017) A cost-based integrated importance measure of system components for preventive maintenance. Reliab Eng Syst Saf 168:98–104
8. Feng Q, Hai X, Huang B, Zuo Z, Ren Y, Sun B, Yang D (2019) An agent-based reliability and performance modeling approach for multistate complex human-machine systems with dynamic behavior. IEEE Access 7:135300–135311
9. Gao K, Peng R, Qu L, Wu S (2020) Jointly optimizing lot sizing and maintenance policy for a production system with two failure modes. Reliab Eng Syst Saf 202:106996

10. Gao X, Cui L, Li J (2007) Analysis for joint importance of components in a coherent system. Eur J Oper Res 182(1):282–299
11. Haddad G, Sandborn PA, Pecht MG (2012) An options approach for decision support of systems with prognostic capabilities. IEEE Trans Reliab 61(4):872–883
12. Iscioglu F (2021) A new approach in the mean residual lifetime evaluation of a multi-state system. Proc Inst Mech Eng, Part O: J Risk Reliab 235(4):700–710
13. Jafary B, Mele A, Fiondella L (2020) Component-based system reliability subject to positive and negative correlation. Reliab Eng Syst Saf 202:107058
14. Levitin G, Finkelstein M, Dai Y (2018) Mission abort policy balancing the uncompleted mission penalty and system loss risk. Reliab Eng Syst Saf 176:194–201
15. Levitin G, Xing L, Xiang Y (2021) Optimal multiple replacement and maintenance scheduling in two-unit systems. Reliab Eng Syst Saf 213:107803
16. Liu Y, Chen Y, Jiang T (2020) Dynamic selective maintenance optimization for multi-state systems over a finite horizon: a deep reinforcement learning approach. Eur J Oper Res 283(1):166–181
17. Peng H, van Houtum GJ (2016) Joint optimization of condition-based maintenance and production lot-sizing. Eur J Oper Res 253(1):94–107
18. Si S, Dui H, Zhao X, Zhang S, Sun S (2012) Integrated importance measure of component states based on loss of system performance. IEEE Trans Reliab 61(1):192–202
19. Sun B, Li Y, Wang Z, Yang D, Ren Y, Feng Q (2021) A combined physics of failure and Bayesian network reliability analysis method for complex electronic systems. Process Saf Environ Prot 148:698–710
20. Tan Z, Li J, Wu Z, Zheng J, He W (2011) An evaluation of maintenance strategy using risk based inspection. Saf Sci 49(6):852–860
21. Wu S, Coolen FP (2013) A cost-based importance measure for system components: an extension of the Birnbaum importance. Eur J Oper Res 225(1):189–195
22. Wu S, Chen Y, Wu Q, Wang Z (2016) Linking component importance to optimisation of preventive maintenance policy. Reliab Eng Syst Saf 146:26–32
23. Wu S, Coolen FP, Liu B (2017) Optimization of maintenance policy under parameter uncertainty using portfolio theory. IISE Trans 49(7):711–721
24. Xiao H, Chen H, Lee LH (2019) An efficient simulation procedure for ranking the top simulated designs in the presence of stochastic constraints. Automatica 103:106–115
25. Xing L (2020) Reliability in internet of things: current status and future perspectives. IEEE Internet Things J 7(8):6704–6721
26. Xu Z, Ramirez-Marquez JE, Liu Y, Xiahou T (2020) A new resilience-based component importance measure for multi-state networks. Reliab Eng Syst Saf 193:106591
27. Yeh WC (2005) A new approach to evaluate reliability of multistate networks under the cost constraint. Omega 33(3):203–209
28. Yuan Z, Khakzad N, Khan F, Amyotte P (2015) Risk-based optimal safety measure allocation for dust explosions. Saf Sci 74:79–92
29. Zhang X, Zeng J (2017) Joint optimization of condition-based opportunistic maintenance and spare parts provisioning policy in multiunit systems. Eur J Oper Res 262(2):479–498

Chapter 5
Importance Measures for Networks

Abstract This chapter focuses on importance measures of networks and prioritises network maintenance and the importance measures of mono-layer networks and multi-layer networks. Regarding mono-layer networks, this chapter analyses the failures of mono-layer network components, including node failures or edge failures, and network failures. The cascading failure process, attack patterns, network failure indicators, load distribution policies, and network failure indicators of monolayer networks are also investigated. Secondly, in multi-layer networks, each layer has a specific network structure, and the number of nodes and the network structure of each layer are different.

Keywords Monolayer network · Multi-layer network · Cascading failure

While existing literature has published a lot of work on the failure analysis of multi-layer networks, few studies have used clustering to describe the change of the topology of a network when it fails. Additionally, most studies on cascading failures have focused on the characteristics of node degree distributions, but they ignore the effect of edges on cascading damage and network stability. The role of internal connections of nodes and edges and their constituent topology have not been studied deeply enough. Therefore, in the analysis of multi-layer network failures, the proposed model integrates different types of nodes and edges and their joint contributions to the network. It also analyses a cluster aggregation-based cascade failure model in cyber-physical systems to describe the network's topological changes during cascade failure through the aggregation of multiple related clusters to complement the existing studies. This chapter also considers the situations under which resources are limited during the maintenance of the network and maintenance decisions need to be introduced to ensure maximum performance improvement with certain resources. That is, this chapter discusses the maintenance decision of the network from the perspective of maintenance priority.

© The Author(s), under exclusive license to Springer Nature Switzerland AG 2024
H. Dui and S. Wu, *Importance-Informed Reliability Engineering*, Springer Series in Reliability Engineering, https://doi.org/10.1007/978-3-031-52455-4_5

5.1 Failure Analysis for Mono-layer Networks

In some cases, a small failure that occurred at the core of a network system may cause the entire network to be paralysed [2, 3]. Some researchers focused on the internal interactions or cascading failure of networks. Rodriguez-Mendez et al. [11] explored the characteristics of the clustering during a cascading failure of a fluid network and its consequence. Zhou et al. [18] analysed the change process of a network cascading failure from the perspective of load dynamics and node dependence. For a clustered supply chain network, Wang and Xiao [12] comprehensively considered the topology of the network, the flexibility of nodes and the efficiency of nodes, and gave a cascading failure model based on an improved ant colony algorithm. Through a simulation method to gradually remove nodes and recalculate the importance in a random network, Li and He [7] measured a small network and a scale-free network, respectively, in which the reliability of networks and the influential nodes are defined in assessing the changes of the networks topology.

5.1.1 Modelling the Mono-layer Network

This section uses complex network theory to model real networks. Complex network theory is widely used to model traffic networks, power grids, communication networks and other networks [4].

A mono-layer network is composed of nodes and edges. The network can be denoted by $\boldsymbol{G} = (\boldsymbol{V}, \boldsymbol{E})$, where $\boldsymbol{V} = \{v_i | i = 1, 2, \ldots, n\}$ represents the set of vertices and $\boldsymbol{E} = \{e_{ij} \mid i, j = 1, 2, \ldots, n\}$ represents the set of edges.

Define the connection relationship between two nodes as $L_{i \to j}$, as shown in Eq. (5.1)

$$L_{i \to j} = \begin{cases} 1, & v_i \text{ and } v_j \text{ have a connection} \\ 0, & v_i \text{ and } v_j \text{ have no connection} \end{cases}. \tag{5.1}$$

Then, the state of the connection between any two nodes can be represented by an adjacency matrix $\boldsymbol{A} = [a_{ij}]_{n \times n}$, where $a_{ij} = 1$ if two nodes are directly connected and $a_{ij} = 0$ otherwise.

In general, each component such as nodes or edges in the network have two states: working and failed. A mono-layer network can fail due to a component failure and a network failure, or a node failure or an edge failure.

5.1.2 Node Failure and Edge Failure

Assume the states of each node and each edge in the network are independent of each other. The components of the mono-layer networks mainly include nodes and edges. Let parameter $\mathcal{S}_{c_i}$ represent the state of components of the mono-layer networks,

$$\mathcal{S}_{c_i} = \begin{cases} 1, \text{ if component } c_i \text{ is working} \\ 0, \text{ if component } c_i \text{ is failed} \end{cases}. \tag{5.2}$$

The states of components can transition. Once a component is successfully maintained, $\mathcal{S}_{c_i}$ changes from 1 to 0. The scale of failure caused by a node failure and an edge failure is different in different networks. Therefore, node failures, edge failures, or the mixture of failures of nodes and edges will be studied according to the characteristics of the network.

5.1.3 Network Failure

Assume a mono-layer network has two states. Set a parameter $\mathcal{S}_G$ to represent the state of the mono-layer networks,

$$\mathcal{S}_G = \begin{cases} 1, \text{ if the network } G \text{ is working} \\ 0, \text{ if the network } G \text{ is failed} \end{cases}. \tag{5.3}$$

When the components of the network are successfully maintained, the network gradually recovers to the working state. $\mathcal{S}_G$ changes from 1 to 0. When most components of the network fail, the network will fail.

Denote the number of components failures by $\mathcal{N}$ and the network efficiency by $\mathbb{E}$, which can be used as metrics to the probability of measure network failure

$$\mathcal{N} = \left|N'\right|, \tag{5.4}$$

and

$$\mathbb{E} = \frac{1}{N(N-1)} \sum_{\substack{i \neq j \\ i,j \in V}} \frac{1}{d_{ij}}, \tag{5.5}$$

where N' is the set of failure nodes, N is the number of nodes, and d_{ij} is the shortest path in the network.

5.1.4 Cascading Failure Analysis for Mono-layer Networks

This section takes node failures as an example to analyse the cascading failures of a mono-layer network, which is based on the load capacity model. We make the following assumptions.

Assumption 1: The load of a node does not exceed its capacity when the node is not failed. When the load exceeds the capacity, the node is failed and its load needs to be redistributed to somewhere else.

Assumption 2: The load of a failed node will only be transferred to the neighbouring normal nodes.

Assumption 3: The capacity of each node is constant.

Under the above assumptions, there are the following modes of a node.

(1) Attack mode. Complex networks often suffer from various kinds of disturbance and attacks such as intentional attacks and unintentional attacks, random attacks and non-random attacks, comprehensive attacks, etc. If the information (e.g., the structure) of a network is available, the network will be deliberately attacked. The most important nodes in the network are the first to be attacked. This will greatly affect the structure of the network. If the information of the network is incomplete, the network will be randomly attacked. This attack mode has little impact on the network than an intentional attack. A comprehensive attack is also a common attack mode. In order to fail a network quickly, a comprehensive attack deliberately attacks important nodes first and randomly attacks the other nodes at the same time.

(2) Initial load of nodes. In many existing studies, the node betweenness or node degree was mostly used as the initial load L_i^0 of nodes. Sometimes researchers will determine the initial load of nodes according to the characteristics of the research network. For example, for the subway network, the passenger flow of each station is generally used as the initial load of each node.

(3) Capacity of nodes. For a real network, the nodes of the network are affected by various factors and their capacity will be limited. For example, the design and construction of subway stations are affected by cost and topography, so each station has a different capacity.

When one studies complex systems, the load capacity limitation of components should be considered. In the load capacity model, the initial load is generally used to determine the node capacity. According to the model proposed by Motter and Lai [9], the capacity, C_i, of node i is proportional to the initial capacity of the node, as shown in Eq. (5.6)

$$C_i = (1 + \alpha)L_i^0, \tag{5.6}$$

where α is the tolerance factor, $\alpha > 0$ and L_i^0 is the initial capacity of node i.

(4) Load distribution strategies. The commonly used load distribution strategies include the average distribution strategy, the closeness distribution strategy, and the node idle capacity distribution strategy.

If node i is attacked and then failed, one can use the network topology information or other information to redistribute the load $\Delta L_{i \to j}$, $(j \in U_i)$ of node i to its neighbouring nodes in a certain proportion. The commonly used load distribution strategies include the average distribution strategy, the closeness distribution strategy, and the node idle capacity distribution strategy.

If the capacity of an adjacent node j can bear its load, the state of node j is normal. Otherwise, the load will continue to be transferred to the neighbouring nodes in a certain proportion. The process of successive failures is called a cascading failure. If the failure of a node is not properly repaired, it can cause more serious damage and eventually lead to the failure of a larger scale. The state of the mono-layer network will be changed from normal to failed. It should be noted that the definition of a cascading failure is different in different disciplines. The cascading failure process is shown in the following.

Step 1: Node i is attacked and is then failed. It is the first failed node in the network.
Step 2: The load of node i will be distributed to the neighbouring nodes.
Step 3: If $C_j > L_j + \Delta L_{i \to j}$ j $\in U_i$, node j is normal. If there are no neighbouring nodes that fail, go to Step 6. Otherwise, if node j is failed, go to Step 4.
Step 4: Count the number of failed nodes.
Step 5: Redistribute the load of node j to its neighbouring nodes. If $C_x > L_x + \Delta L_{j \to x}$, node x is normal. Go to Step 5. Otherwise, node x is failed. Go to Step 4.
Step 6: There is no new failed nodes appear in the network. The cascading failure ends.

5.2 Failure Analysis for Multi-layer Networks

Multi-layer networks are connected by complex networks or complex systems. Existing research has shifted from the analysis of monolayer networks to the analysis of multi-layer networks. In related works, existing studies tend to use clustering to deal with the analysis process of topologies that contain two or more layers of networks, while the clustering phenomenon that occurs in monolayer networks has not been properly explored. To some extent, the clustering phenomenon in monolayer networks can be considered as a special case of multi-layer networks. In addition, little study has used the method of clustering aggregation to characterise the changes in topology when a particular network fails. This section focuses on this issue and proposes a cascading failure model based on clustering aggregation, while giving a quantitative approach to characterising the key indicators that affect the failure process.

5.2.1 Related Work

In terms of the network performance and reliability, Xing and Dugan [16] proposed a generalized phased-mission system (GPMS) analysis methodology called GPMS-CPR to analyse the reliability, performance, and sensitivity of GPMS. Levitin and Dai [5] introduced service reliability and performance indices and presented a fast numerical algorithm for their evaluation for arbitrary subtask distributions in a grid with star architecture. Levitin and Xing [6] presented an algorithm for evaluating the performance distribution of complex series-parallel multi-state systems with common cause failures due to the propagation of failures of components in a network. Booker et al. [1] showed how the performance of a cellular network during and immediately after future hurricanes can be estimated based on a combination of hurricane wind field models, structural reliability analysis, Monte Carlo simulation, and cellular network models and simulation tools. Wu et al. [15] studied the performance of metro networks from a network science perspective. Wu and Baker [14] proposed a statistical learning technique (i.e., random forests) to efficiently estimate network performance in place of direct physical simulation. Niu [10] developed an algorithm iteratively separates capacity vectors that satisfy the required capacity level from the universal space to measure the performance of a multi-state flow network. Wang et al. [13] modelled the high-speed railway as a three-layer network including topological, functional, and service layers and assessed the integrated network performance from the view of transportation accessibility. Zhang et al. [17] proposed a cascading reliability model to model, measure, and control coupling performances against cascading failures. Ma et al. [8] presented a technique for the probabilistic simulation of power transmission systems under hurricanes and provides fundamental insights on the modelling and quantification of power system performance and resilience.

A failure caused by one or a few nodes or connections in a network can cause other nodes to fail through the coupling relationship between nodes, eventually leading to the failure of a larger range of nodes or the collapse of the entire network, which is the cascading failure model described in Sect. 5.1.4. In this section, the cascading failure model is extended to the multilayer network. The extended cascading failure model is based on clustering aggregation, which quantifies the heterogeneous locations of network elements by giving metrics such as classified nodes, classified clusters, the degree of node irregularity, and the degree of cluster irregularity. Furthermore, this section classifies and quantifies the general network topology in a hierarchical manner, dividing the failure process into two phases: intra-cluster transmission and inter-cluster transmission.

5.2.2 *Classified Nodes*

In this section, classified nodes are defined in terms of the ability of each node to accommodate external load and simultaneously maintain its stability. The level of the node irregularity is a quantitative description of this capability. We present three classification nodes and then elaborate their characteristics and the level of the irregularity of these nodes. The central nodes always have larger connection values and are also an important factor affecting the robustness of a given network in case of deliberate attacks. In a directed network, the outward degree of a node can be considered as an authority feature, which determines the degree of influence of subsequent fault transmission. In contrast, we consider the inward degree of a node as a sink characteristic, which describes the contribution of each node to the stability of the network system. As for the hub characteristic, it plays an important role in the transmission of failures. The values of the inward and outward degree and their quantitative relationships are the three basic characteristics of each node. We focus on the value of $\frac{k_{i,\text{out}}}{k_{i,\text{in}}}$ and all nodes can be quantified; and then obtain three types of nodes in light with the numerical features, namely source nodes, central nodes and convergence nodes, which correspond to sources, hubs and sinks, respectively. For simplicity, we denote δ^1, δ^2 and δ^3 as the three above-mentioned classified nodes, and δ^i $(i = 1, 2, 3)$ represents the ratio of the outward degree to the inward degree for any node. The specific calculation equation is shown in Eq. (5.7).

$$\delta^i = \frac{k_{i,\text{out}}}{k_{i,\text{in}}}. \tag{5.7}$$

The level of irregularity of a node (LIN) is defined as *the ability of a node to maintain itself without failure while receiving external unstable loads*. In general, the value of the node outward degree increases or decreases, and the residual load of each node increases. Therefore, to further investigate how the quantitative relationship between the inward and outward degrees of nodes affects the ability of nodes to absorb external loads, the numerical equation of LIN is defined as follows

$$\begin{aligned} N_i^{\text{lin}} &= \frac{k_{i,\text{out}}}{k_{i,\text{in}}} \\ &= \frac{\sum_{k=0}^{+\infty} k P_i(j,k)}{\sum_{j=0}^{+\infty} j P_i(j,k)}. \end{aligned} \tag{5.8}$$

In a randomly generated directed network model, $\sum_{k=0}^{+\infty} k P_i(j,k)$ is the average of the outward degrees of any node considering different probability cases (the value is $k_{i,\text{out}}$ when it is not a random network). Similarly, $\sum_{j=0}^{+\infty} j P_i(j,k)$ represents the average of the inward degrees of any node ($k_{i,\text{in}}$ for non-random networks). The definition of LIN shows that it reflects the degree of load concentration and the choice of transmission paths, and the load balancing in the network. Higher degrees of outgoing and incoming usually improve the load capacity, transmission efficiency and load balancing of the network. When the inward degree of a node is the same as the outward degree, the node is considered to be in equilibrium and its ability to absorb or reject external loads is reduced to a minimum value of 0. If the two do not agree, the node is not in such a stable equilibrium. For a node, when the outward degree is greater than the inward degree, there is always a tendency to absorb external loads. On the contrary, when the larger the inward degree is, the stronger its tendency to reject the self-load is. It can be seen that the number of nodes with a LIN value of 0 in the network is generally high, which has a passive impact on maintaining the system stability when cascading failures occur.

5.2.3 Classified Clusters

A cluster is a collection of node members with similar characteristics, and classified clusters are a cluster classification that considers the topological characteristics of the clusters. The Level of Irregularity of a Cluster (LIC) describes the load-absorbing capacity of a given cluster, similar to the process of defining LIN. Cluster aggregation is essentially a process in which the set of nodes keeps shrinking due to their similar capabilities, as explained below.

In generic networks, applying clustering to deal with cascading faults has two advantages. Clustering can describe the hierarchical topological characteristics of general networks. Due to the time and space delays in the load propagation process, the nodes after multiple aggregations always form regional characteristics in distribution, and the load absorption capacity of nodes in the upstream region is always stronger than that in the downstream region. Therefore, the modularity makes the nodes closely correlated with each other, thus forming clustering groups at different levels, which is meaningful for analysing how failure paths are formed. The introduction of the cluster concept simplifies the analysis process of the potential propagation energy distribution characteristics. The load propagation process is divided into intra-cluster transfer and inter-cluster transfer. This treatment can distinguish the differences between the two propagation methods and selectively ignore unnecessary factors.

The LIC is based on the definition of LIN, which describes the ability of the group of nodes as a whole to withstand external unstable loads, and it is therefore essentially a similarity ratio with its definition shown in Eq. (5.9)

$$N_i^{\text{lic}} = \frac{\sum_{i=1}^{N_H} \sum_{k=0}^{+\infty} k P_i^H(j,k)}{\sum_{i=1}^{N_H} \sum_{j=0}^{+\infty} j P_i^H(j,k)}. \tag{5.9}$$

Clusters consisting of multiple nodes have the same characteristics as those of simple nodes. When determining the inward and outward degrees of a certain cluster, the principle of ignoring the edges between internal nodes should be followed first. In other words, we should consider only the distribution of edges located in different clusters. In Eq. (5.9), $\sum_{i=1}^{N_H} \sum_{k=0}^{+\infty} k P_i^H(j,k)$ is the outward degree of a cluster, which is essentially equal to the sum of the outward degrees of the edges connected to other clusters. Similarly, $\sum_{i=1}^{N_H} \sum_{j=0}^{+\infty} j P_i^H(j,k)$ denotes the inward degree of the same cluster. The LIC of a cluster is influenced by the topology, the common composition of the node members and the degrees of the distribution of these nodes. When the inward and outward degrees of a cluster are the same, the LIC with the minimum cluster, indicates that the cluster is less capable of absorbing or shifting the residual load.

We define classification clusters according to different LIC values. Similar to the sorting of classification nodes, classification clusters can be classified into three types: source, central, and convergence. Source clusters do not contain inward degree edges and have more outward degree edges, so the LIC of the source clusters is always larger. Centre clusters have both inward degree edges and outward degree edges and are always located at the centre of the network. Their LIC is related to the ratio of inward degrees to outward degrees in the same cluster. Convergence clusters refer to a group of nodes that contain only inward degree edges. Since the load can only enter the convergence cluster, the remaining space for the convergence clusters to absorb the external load is gradually reduced, so the LIC of the convergence clusters is always smaller.

(1) As for source clusters, we obtain Eqs. (5.10) and (5.11)

$$k_{\text{in}}^{Cl} = 0, \tag{5.10}$$

and

$$\begin{aligned} k_{\text{out}}^{Cl} &= \sum_{i=1}^{N_A} k_{\text{out}}^{Cl} (\text{or}) k_{i,\text{out}}^{Cl} \\ &= \sum_{i=1}^{N_A} k_{i,\text{out}}^{Cl} + \sum_{i=1}^{N_H} k_{i,\text{out}}^{Cl}. \end{aligned} \tag{5.11}$$

From Eqs. (5.10) and (5.11), the LIC of the source cluster can be calculated through $N^{\text{LIC}^{Cl-1}} = \frac{k_{\text{out}}^{Cl}}{k_{\text{in}}^{Cl}}$, ($\text{LIC}^{Cl-1} \to +\infty$) and in Eq. (5.11), when the source cluster does not contain a central node, its outward degree is determined by the sum of the outward degrees of all source nodes belonging to the same source cluster. Otherwise, the outward degree of a node is influenced by the sum of the outward degrees of all source nodes and the sum of the outward degrees of the corresponding central nodes. By the way, we selectively ignore $\sum_{i=1}^{N_H} k_{i,\text{in}}^{Cl}$ of all central nodes and consider them as inner edges connecting some nodes. The inward degree of the cluster is the same as the sum of the inward degrees of all source nodes, both being 0. Therefore, the LIC of the source cluster is infinite.

(2) As for the central cluster, we obtain Eqs. (5.12) and (5.13)

$$\begin{aligned} k_{\text{in}}^{Cl} &= \sum_{i=1}^{N_A} k_{i,\text{out}}^{Cl} + \sum_{i=1}^{N_H} k_{i,\text{in}}^{Cl} (\text{or}) k_{\text{in}}^{Cl} \\ &= \sum_{i=1}^{N_A} k_{i,\text{out}}^{Cl} + \sum_{i=1}^{N_H} k_{i,\text{out}}^{Cl}, \end{aligned} \tag{5.12}$$

and

$$\begin{aligned} k_{\text{out}}^{Cl} &= \sum_{i=1}^{N_H} k_{i,\text{out}}^{Cl} + \sum_{i=1}^{N_S} k_{i,\text{out}}^{Cl} (\text{or}) k_{\text{out}}^{Cl} \\ &= \sum_{i=1}^{N_H} k_{i,\text{out}}^{Cl} + \sum_{i=1}^{N_S} k_{i,\text{in}}^{Cl}. \end{aligned} \tag{5.13}$$

We use the same method to measure the central cluster through $N^{\text{LIC}^{Cl-2}} = \frac{k_{\text{out}}^{Cl}}{k_{\text{in}}^{Cl}}$, $(1 < \text{LIC}^{Cl-2} < +\infty)$, and in Eq. (5.13). If there are no central node members in the source cluster, the inward degree of the central cluster is determined by the sum of the outward degrees of all source nodes and the corresponding inward degrees of some central nodes, or the inward degree of the cluster is determined by the sum of the outward degrees of all source nodes and the sum of the outward degrees of these central nodes. In Eq. (5.7), again, if there is no central node in the convergent cluster, the outward degree of the central cluster is equal to the sum of the outward degrees of all convergent nodes and the outward degrees of some central nodes, or the outward degree of the central cluster is determined by the sum of the inward degrees of all convergent nodes and the outward degrees of the same central nodes.

(3) The last one is the convergence cluster, which is concluded through (5.14) and (5.15)

$$\begin{aligned} k_{\text{in}}^{Cl} &= \sum_{i=1}^{N_S} k_{i,\text{in}}^{Cl} (\text{or}) k_{\text{in}}^{Cl} \\ &= \sum_{i=1}^{N_S} k_{i,\text{in}}^{Cl} + \sum_{i=1}^{N_H} k_{i,\text{in}}^{Cl}, \end{aligned} \tag{5.14}$$

and

$$k_{\text{out}}^{Cl} = 0. \tag{5.15}$$

Through Eqs. (5.14) and (5.15) and $N^{\text{LIC}^{Cl-3}} = \frac{k_{\text{out}}^{Cl}}{k_{\text{in}}^{Cl}}$, $(\text{LIC}^{Cl-3} \approx 0)$, we can analyse the LIC of convergent clusters. In Eq. (5.14), when a convergence cluster does not contain any central node, its inward degree is the sum of the inward degrees of all convergence nodes. Otherwise, the value is equal to the sum of the inward degrees of all convergence nodes and the sum of the inward degrees of the corresponding central nodes.

5.2.4 *Relative Circulation Indicators*

In the previous section of classified nodes and classified clusters, the definition, interpretation, and clustering aggregation analysis of classification nodes and classification clusters were presented in detail. This section starts from analysing the degree distribution characteristics of nodes in order to further distinguish the metrics of categorical nodes and those of categorical clusters. We therefore propose three relative circulation metrics, namely, relative circulation level, relative load transfer rate and relative edge traffic capacity, based on which the circulation difficulty of different edges are described. As for the relative circulation level, it depends on the degree of difference between the two ratios. One is the ratio of the outward degree edges of upstream nodes that connect downstream nodes to the total outward degree edges. The other is the ratio of inward degree edges to the total inward degree edges of the upstream node that connects to the downstream node. The smaller the difference, the smoother the load flow between the two nodes. These two nodes belong to each other's key upstream and downstream nodes. Therefore, RLC is a static property that measures the continuity of two nodes, which is quantitatively described as follows

$$\begin{aligned} N_{ij}^{\text{rlc}} &= \frac{1}{\left|\frac{1}{k_{i,\text{out}}} - \frac{1}{k_{j,\text{in}}}\right|} \\ &= \frac{k_{i,\text{out}} k_{j,\text{in}}}{\left|k_{j,\text{in}} - k_{i,\text{out}}\right|}. \end{aligned} \tag{5.16}$$

In Eq. (5.16), $\frac{1}{k_{i,\text{out}}}$ and $\frac{1}{k_{i,\text{in}}}$ represent the two ratios mentioned above, and node i and node j represent the upstream node and downstream node, respectively. A large difference between $\frac{1}{k_{i,\text{out}}}$ and $\frac{1}{k_{i,\text{in}}}$ suggests a strong correlation between the two nodes and easier to shift the tide from node i to node j.

Regarding the relative load transfer rate, we can derive it from another metric, namely the relative load transfer. This metric is used to calculate the relative amount of load received by the downstream node when transferring each unit load from the upstream node. Thus, the relative load transfer rate is the difference between the relative load transfer rates and is used to describe the relative amount of load transfer per unit of time. Finally, we propose the following equations to measure and interpret this metric

$$\begin{aligned} N_{ij}^{\text{rlt}}(t) &= \frac{Q_j(t)}{Q_i(t)} \\ N_{ij}^{\text{rtr}}(t) &= \frac{d}{dt}\left[RLT_{ij}(t)\right] \\ &= \frac{d}{dt}\left[\frac{Q_j(t)}{Q_i(t)}\right] \end{aligned}$$

$$
\begin{aligned}
&= \frac{k_i Q_j^0 + k_j Q_i^0}{(Q_i^0 - k_i t)^2}. \\
Q_i(t) &= Q_i^0 - k_i t \\
&= Q_i^0 - \frac{k_{i,\text{out}}}{k_{i,\text{in}}} t,\ Q_j(t) \\
&= Q_j^0 + k_j t \\
&= Q_j^0 + \frac{k_{j,\text{out}}}{k_{j,\text{in}}} t.
\end{aligned} \tag{5.17}
$$

In Eq. (5.17), $Q_i(t)$ and $Q_j(t)$ represent the real-time load capacity of upstream node i and downstream node j, respectively, as a function of time. The ratio of $Q_i(t)$ and $Q_j(t)$ can be used to measure the relative load transfer rate. Q_i^0 and Q_j^0 are the initial loads of node i and node j. $\frac{k_{i,\text{out}}}{k_{i,\text{in}}}$ and $\frac{k_{j,\text{out}}}{k_{j,\text{in}}}$ denote the absolute values of transmission loads of node i and node j. RLT represents the ratio of the real-time load capacity of the downstream node to that of the upstream node, i.e., the load capacity specific load. It reflects the load level of the downstream node with respect to the upstream node. RTR is the derivative of RLT and represents the rate of change of the load capacity ratio.

When comprehensively considering RLC, RTR, their difference and intrinsic connection, the relative flow capacity of an edge, or RCF, can be obtained as an indicator of the load transfer capacity of a particular edge. Essentially, it is numerically equal to the maximum amount of load per unit of time allowed to pass through the edge we explore. Moreover, the RCF depends on the static environment and on the real-time load flow. That is, it is determined by the static factor RLC and the dynamic factor R, respectively. Similar to the flow equation, we give Eq. (5.18) to obtain the RCF

$$
N_{ij}^{\text{rcf}}(t) = N_{ij}^{\text{rlc}}(t) * N_{ij}^{\text{rtr}}(t). \tag{5.18}
$$

It can be seen that the maximum load allowed per unit of time between any upstream node i and downstream node j is closely related to both the real-time transmission load and the relative circulation level between the corresponding nodes.

5.2.5 *Cascading Failure Models in a Special Multi-layer Network*

In this section, we construct a cascading failure model based on cluster aggregation. First, six basic assumptions for building the model are given. Secondly, the rationality of cluster aggregation propagation is verified by the proof of quantitative indexes. Finally, based on the hierarchical and cumulative effects of cluster aggregation, a failure transmission model is constructed, which regulates the failure load transmission process alternately according to the transmission order between clusters and clusters.

5.2.5.1 Basic Hypothesis

We make the following assumptions.

Hypothesis 1: Regardless of the case where the node load exceeds the maximum capacity and the node has not failed. In other words, the node will fail once the load exceeds its capacity threshold.

Hypothesis 2: Assume that when a node fails, all loads are transferred to the adjacent non-failed node, ignoring the phenomenon of load transfer ratio and residual load. That is, after the node fails, the residual load is 0.

Hypothesis 3: The influence of the construction cost factor of a node on the bearing capacity is not considered.

Hypothesis 4: There are always other nodes when the initial node fails, and it is assumed that all node crashes due to partial initial failure nodes only occur at least 2 or more times during the failure process.

Hypothesis 5: Assume that the capacity threshold of a node is constant and does not consider fluctuations within a certain range.

Hypothesis 6: Random attacks and intentional attacks on nodes or edges are not considered simultaneously; the model is only based on the premise of considering random attacks.

5.2.5.2 Rationality of Cluster Aggregation

In this subsection, we explain the rationality of building a cascading failure model based on cluster aggregation, which is divided into two parts. First, the cumulative effect in general networks is the basis for explaining the characteristics of cascades in networks. The first subsection will demonstrate the reason for the cumulative effect. Second, the cascading effects in the network are the basis for applying cluster generation as a tool to analyse fault processes, and the second subsection will explain some specific manifestations of the cascading features.

As the network operates, the residual load storage capacity of downstream nodes gradually decreases, which means that it is more difficult for them to keep their functions stable while absorbing the residual load. To quantitatively describe this phenomenon, we propose to assemble a series of clusters containing nodes at different levels $H_1, H_2, \ldots H_a, \ldots,$ where each cluster H_a is a set containing a similar set of nodes denoted as $\left\{H_a^1, H_a^2, \ldots, H_a^{n_a}\right\}$. We assume that i can denote any node at any level and the cumulative effect can be represented by the following process.

$$
\begin{aligned}
Q(H_2^i) &= \sum_{k=1}^{k_{i,\text{in}}^2} \left[Q(H_1^i)\eta(H_1^i)\right], i \in (1, n_2)\\
Q(H_3^i) &= \sum_{k=1}^{k_{i,\text{in}}^3} \left[Q(H_1^i)\eta(H_1^i) + Q(H_2^i)\eta(H_2^i)\right], i \in (1, n_3)\\
&\vdots\\
Q(H_{a+1}^i) &= \sum_{k=1}^{k_{i,\text{in}}^{a+1}} \left[Q(H_1^i)\eta(H_1^i) + Q(H_2^i)\eta(H_2^i) + \cdots + Q(H_a^i)\eta(H_a^i)\right]\\
&= \sum_{k=1}^{k_{i,\text{in}}^{a+1}} \sum_{a=1}^{a} \left[Q(H_a^i)\eta(H_a^i)\right], i \in (1, n_{a+1}).
\end{aligned}
\tag{5.19}
$$

In Eq. (5.19), $Q(H_{a+1}^i)$ is the absorption function of cluster H_{a+1}^i facing the residual load of the upstream cluster at level H_{a+1}. $\eta(H_a^i)$ is the efficiency function of cluster H_a^i at level H_a. $k_{i,\text{in}}^{a+1}$ is the inward degree of any cluster at any level.

It is obvious that $Q(H_{a+1}^i) \geq 0$ is a constant establishment, so $Q(H_2^i) < Q(H_3^i) < \cdots < Q(H_{a+1}^i)$ is reasonable. That is, for the general network processed by clustering and aggregation in multiple cascading failures, the nodes and edges in the downstream region are more likely to reach their capacity threshold and the threshold distribution is stepped. The downstream nodes and edges are more vulnerable to attack when cascading failures occur.

The level effect is essentially an intuitive description of the load spreading in a stepwise manner during a failure. Due to the cumulative effect, different sequential nodes are exposed to different residual loads. In a network, there is more than one chain, which means that each node has the probability of traversing through different chains since each node always has a unique sequence value. We assume that the amount of nodes in each chain is $T_1, T_2, \ldots, T_a, \ldots, T_{m_i}$. To some extent, the sequence value of every node can be used to measure the topology environment and load flow environment surrounding them. Together all of these sequence values form a hierarchical sequence, which can be expressed by $H_1, H_2, \ldots, H_a, \ldots, H_{m_i}$. As for the level effect of nodes, we define it through the following equations and they are $\overline{H_i} = \frac{1}{m_i}\sum_{a\in m_i} H_a$ and $\overline{T_i} = \frac{1}{m_i}\sum_{a\in m_i} T_a$. Among the two equations, $\overline{H_i}$ represents the average level value of node i and $\overline{T_i}$ for the average sequence value of node i. As for $\forall i, j \in n$, if $\exists \epsilon_0$ is small enough and makes the inequality $\left|\frac{\overline{H_i}}{\overline{T_i}} - \frac{\overline{H_j}}{\overline{T_j}}\right| \leq \epsilon_0$ hold, we regulate that nodes i and j belong to the same level. When we gather all nodes that have similar properties together, a set is formed and we call it the cluster.

The horizontal effect of the cluster leads to an increase in the difficulty of the node to absorb the residual load or to be traversed by the tide, i.e., the propagation potential energy of the node becomes weaker. The propagation potential energy of a node is defined as follows. In contrast to the sandpile theory, a general network is a dynamic system consisting of a series of node elements with different propagation potential energies. The dynamic system composed of elements with different propagation potential energy, all nodes have the tendency to keep the system in the state with the lowest network-wide propagation potential energy, a state that is favourable to face another random attack. Further, cascading failures makes the finding of this phenomenon even more compelling. First, the propagation potential energy of a single node is an appropriate description of the storage's ability to absorb the remaining load. The process of potential energy change between nodes can be reflected by the edges of connected nodes. The propagation potential energy of upstream nodes is always greater than that of downstream nodes. That is, in a directed network, the propagation potential energy at the end of the arrow is always greater than that at the front of the arrow, as a result of the cumulative effect. In order to find the quantitative correlation between the LIN of any node and the topological environment where the node is located, we describe it from two perspectives.

First of all, for node j during interval $(0, t)$, load transferring the node can be divided into outflow load $Q_{j,\text{out}}(t)$ and inflow load $Q_{j,\text{in}}(t)$. Taking outflow load as an example, it is affected by a collection of downstream nodes adjacent to node j, which can be described as $K = \{k_1, k_2, \ldots, k_{j,\text{out}}\}$, while the LIN of these downstream nodes also heterogeneously absorb the residual load from node j, so the total amount of outflow load is $Q_{j,\text{out}}(t) = \sum_{k=1}^{k_{j,\text{out}}} \frac{k_{k,\text{out}}}{k_{k,\text{in}}} * t$. By analogy, we gain the collection of upstream nodes adjacent to node j, which is shown through the set $I = \{i_1, i_2, \ldots, i_{j,\text{in}}\}$ and similarly the total amount of inflow load is $Q_{j,\text{out}}(t) = \sum_{k=1}^{k_{j,\text{out}}} \frac{k_{k,\text{out}}}{k_{k,\text{in}}} * t$. Thus, we use a new amount of outflow of load per unit of time $Q'_{j,\text{out}}(t) - Q'_{j,\text{in}}(t)$ to measure the node's ability to absorb load or be traversed by other nodes. After simplification, we obtain Eq. (5.20)

$$\begin{aligned}
\frac{d\left(Q_{j,\text{out}}(t) - Q_{j,\text{in}}(t)\right)}{dt} &= \frac{d}{dt}\left(\sum_{k=1}^{j,out} \frac{k_{k,\text{out}}}{k_{k,\text{in}}} t\right) - \frac{d}{dt}\left(\sum_{i=1}^{jin} \frac{k_{i,\text{out}}}{k_{i,\text{in}}} t\right) \\
&= \sum_{k=1}^{k_{j,\text{out}}} \frac{k_{k,\text{out}}}{k_{k,\text{in}}} - \sum_{i=1}^{k_{j,\text{in}}} \frac{k_{i,\text{out}}}{k_{i,\text{in}}} \\
&= k_{j,\text{out}} \overline{\frac{k_{k,\text{out}}}{k_{k,\text{in}}}} - k_{j,\text{in}} \overline{\frac{k_{i,\text{out}}}{k_{i,\text{in}}}}.
\end{aligned} \tag{5.20}$$

On the other hand, the capacity of node j to absorb external loads is related to its inward and outward degrees. Therefore, the new load inflow per unit of time corresponds to the ratio of inward and outward degrees, and Eq. (5.20) can be simplified as $\frac{k_{j,\text{out}}}{k_{j,\text{in}}} = k_{j,\text{out}} * \overline{\frac{k_{k,\text{out}}}{k_{k,\text{in}}}} - k_{j,\text{in}} * \overline{\frac{k_{i,\text{out}}}{k_{i,\text{in}}}}$, where $\overline{\frac{k_{k,\text{out}}}{k_{k,\text{in}}}}$ and $\overline{\frac{k_{i,\text{out}}}{k_{i,\text{in}}}}$ can be considered as the storage capacity factors of node j in the downstream and upstream regions, respectively. We denote them as a and b and proceed to simplify Eq. (5.20). Then we obtain $k_{j,\text{out}} = \frac{b}{a}(k_{j,\text{in}} - \frac{1}{a}) + \frac{1}{a^3(k_{j,\text{in}} - \frac{1}{a})} + \frac{2}{a^2}$. It can be seen that the value of $k_{j,\text{out}}$

is minimized when $ak_{j,\text{in}} = 1$. In this case, the capacity of node j to absorb external loads reaches its lower limit $\frac{2(\sqrt{b}+1)}{a}$, which is strongly related to its surroundings.

5.2.6 Construction of Failure Model

In this subsection, a cascading fault model based on clustering aggregation is built. The model is divided into six phases and includes three judgment conditions. One is whether the initial fault node belongs to one of the clusters of edge nodes, whose edges always connect different clusters. Another one is whether the downstream nodes belong to the same cluster member. The last one is when the downstream nodes belong to the same cluster, we explore whether there are multiple edges connected to nodes belonging to the same cluster. These six phases are described in detail below.

Phase I: Any node i in the network may face random attacks. L_i is the real-time load of node i and C_i is the maximum load-absorbing capacity of node i. If $L_i \leq C_i$ is specified, then the node can maintain its functionality and will not fail. Otherwise, node i will fail and L_i is the remaining load to be passed to downstream nodes and edges. We move to phase II.

Phase II: We start checking whether node i belongs to one of the sets of edged nodes, and if so goes to phase III, or to phase VI. Edged nodes are always associated with different clusters, and those without edged nodes only have edges connecting nodes of the same cluster.

Phase III: We continue to analyse the location characteristics of each node in case of cascading failure. When node i is connected to a downstream node through only one edge, i.e., $k_{i,\text{out}} = 1$ or more nodes, which are both located in the same cluster, i.e., $k_{i,\text{out}} > 1$ and $k_j = \{k_{j_1}, k_{j_2}, \ldots, k_{j_n}\}$, where k_j denotes the set of downstream nodes, then it enters phase V. If not, then it enters phase IV.

Phase IV: Considering that the failed node is one of the edge nodes and its downstream nodes are not in the same cluster, we define this process as the transfer phase between clusters, and we use the following load transfer scenario to describe the load transfer process

$$D(i, j_n) = \begin{cases} \text{rank}\{N_{ij_n}^{\text{rcf}}\}, & \text{if } N_{ij_1}^{\text{lic}} = N_{ij_2}^{\text{lic}} = \cdots = N_{ij_n}^{\text{lic}} \\ \text{rank}\{N_{ij_n}^{\text{rcf}}(\max N_{ij_n}^{\text{rcf}})\}, & \text{if } \exists N_{ij_c}^{\text{lic}} \neq N_{ij_d}^{\text{lic}} \end{cases}. \tag{5.21}$$

In Eq. (5.21), $D(i, j_n)$ represents a load transfer scenario which is affected by the LIC of the cluster and is limited by the load-transferring capacity of related edges. $N_{ij_n}^{\text{lic}}$ is a set of LIC of each downstream cluster which contains j_n, one of these downstream nodes that connect node i, $\text{rank}\{N_{ij_n}^{\text{rcf}}\}$ represents in the case that the LIC of downstream clusters is equal to each other, we rank and descend the set and then the direction nodes of transfer load is ordered and optimised, only considering the Relative Level of Circulation (RLC) of all related edges, in other words, the load-transferring ability of these edges. More specifically, the cluster with the largest RLC is considered to be transmitted. If this cluster can accommodate the load to be distributed, the load will not flow to another cluster with the next largest RLC. If not, the residual cluster would absorb the load in order, until the remaining load is completely transferred and these clusters can accommodate all remaining load. All these mean the end of a cascading failure. In another case when $N_{ij_c}^{\text{lic}} \neq N_{ij_d}^{\text{lic}}$ is formed, we first descend and rank the LIC of all related clusters. After that, $\text{rank}\{N_{ij_n}^{\text{rcf}}(\max N_{ij_n}^{\text{rcf}})\}$ means on the basis of selecting out the cluster having the largest value and RCF represents the Relative Capacity of the Flow, we begin to rank the RLC of edges that links node i with all related clusters, through which the direction of remaining transfer load is determined.

Phase V: In this case, a load-transferring process occurs between nodes that consist of the same cluster when the initially failed node has only one downstream node connected or when multiple downstream nodes of the initially failed node belong to the same cluster. What's more, this process is based on the LIN of related nodes and we obtain a load distribution scenario as follows

$$d(i, j_n) = \begin{cases} \text{rank}\{N_{ij_n}^{\text{rcf}}\}, & \text{if } N_{ij_1}^{\text{lin}} = N_{ij_2}^{\text{lin}} = \cdots = N_{ij_n}^{\text{lin}} \\ \text{rank}\{N_{ij_n}^{\text{rcf}}(\max N_{ij_n}^{\text{lin}})\}, & \text{if } \exists N_{ij_c}^{\text{lin}} \neq N_{ij_d}^{\text{lin}} \end{cases}. \tag{5.22}$$

In Eq. (5.22), $d(i, j_n)$ represents how load selectively transfers from node i to node j_n. The transfer path also has a strong correlation with the LIN of each downstream node and the load-transferring ability of related edges. $N_{ij_n}^{\text{lin}}$ is the LIN of all downstream nodes j_n and $\text{rank}\{N_{ij_n}^{\text{rcf}}\}$ represents the rank of the RCF for all related edges when the LIN of all downstream nodes are the same. After that, the transfer path of the residual load is determined. Conversely, if there are at least two different LIN values among downstream nodes, in other words, $\exists N_{ij_c}^{\text{lin}} \neq N_{ij_d}^{\text{lin}}$ exists, we first rank and descend the LIN of all downstream nodes and select out the largest one represented by $\max\{N_{ij_n}^{\text{lin}}\}$. Then we again rank and descend nodes with the maximum value of LIN in each cluster and $\text{rank}\{N_{ij_n}^{\text{rcf}}(\max N_{ij_n}^{\text{lin}})\}$ describe such process.

Phase VI: After experiencing Phase II and Phase V, node i does not belong to the set of edged nodes of corresponding clusters. So in this phase load transfers between nodes in the same clusters, which is similar to Phase V and we check that if the real-time load, including external load and initial load, exceeds the upper threshold of each node. That's to say, if $L_j^0 + \Delta L_{ij_1} \leq C_{j_1}$ is formed, the node could keep its function and cascading failure ends. While if $L_j^0 + \Delta L_{ij_1} > C_{j_1}$ exists, we turn to phase II and the failure would go on until $L_{j_n}^0 + \sum_{j_n \in J} \Delta L_{ij_n} \leq C_{j_n}$ is formed. $J = \{j_1, j_2, \ldots, j_n\}$ is the set of all downstream nodes adjacent to node i and $\sum_{j_n \in J} \Delta L_{ij_n} = L_i^0$ reveals that the sum of transfer load of those which absorb a part amount of load originated from node i is equal to the initial amount of load of node i, which is marked by L_i^0. Furthermore, the set J contains all failed nodes and $i \to j_1 \to j_2 \to \cdots \to j_n$ consists of transfer the path during the failure process, which is the symbol of the end of a certain cascading failure.

5.3 Maintenance Priority Importance for Networks

In this section, we investigate node maintenance and edge maintenance, as well as their cooperation in maintenance. By analysing the factors that affect node maintenance and edge maintenance, we obtain their maintenance policies. When repairing, we should not only pay attention to the failed node, but also to the upstream and downstream nodes connected to the failed node.

5.3.1 Node Maintenance

The maintenance of nodes is affected by the following factors: node degree, initial load of the node itself, the initial load, and degree for the upstream node set, initial load and degree for the downstream node set, and time from node failure to the end of the failure cascade. After the node is repaired, its load and capacity may change. The capacity of a node (NC) can be expressed as a specific performance index under a specific environment in the network index, which is quantified by the following equation.

$$N_i^{nc} = \sum_{q=0}^{k_i^{in}} \frac{Q_i^0 + k_i * T^i}{Q_{p_q}^0 - k_{p_q} * T^i} - \sum_{q=0}^{k_i^{out}} \frac{Q_{r_q}^0 + k_{r_q} * T^i}{Q_i^0 - k_i * T^i}. \tag{5.23}$$

Different nodes have different maintenance methods. Node degree refers to the degree to which the network can be restored to its original operating efficiency under the specified maintenance conditions and within the specified maintenance time. The node degree is expressed in Eq. (5.24).

$$N_i^{nms}(T^i) = \sum_{q=0}^{k_i^{in}} \frac{k_i * Q_{p_q}^0 + k_{p_q} * Q_i^0}{(Q_{p_q}^0 - k_{p_q} * T^i)^2} - \sum_{q=0}^{k_i^{out}} \frac{k_i * Q_{r_q}^0 + k_{r_q} * Q_i^0}{(Q_i^0 - k_i * T^i)^2}, \tag{5.24}$$

where T^i represents the period from the beginning to the end of the cascading failure process of node i. Since node i may have multiple upstream and downstream nodes, k_i^{out} and k_i^{in} represent the out-degree and in-degree of node i, respectively. p_q and r_q represents the upstream node set and the downstream set of the node, respectively. $Q_{p_q}^0$ and $Q_{r_q}^0$ represent the capacity of any node in the upstream node set and the downstream node set, respectively. N_i^{nms} represents the maintenance policy of node i.

Generally, three facts determine the load transfer state of any node in the transportation network: the change in characteristics of the original load of the node itself over time, the change in characteristics of the external load of the upstream node over time, and the change of external load of the downstream node over time. In the cascading failure process caused by a node failure, the load transfer efficiency of different nodes continuously changes over time. If the transfer efficiency changes slowly, the node is maintaining its own stability within a certain period. The range of transfer efficiency that can be tolerated is small. On the contrary, if the transfer efficiency changes faster, the range of transfer rate that can be tolerated by the node is large. When repairing a node upon failures, the time degradation of each node in the network can only reflect the size of the node's ability to affect other nodes in different situations. If the degeneration time of the node is fast, the time from the beginning to the end of the cascading failure is also fast, which suggests that the node has less influence on other nodes. Conversely, it means that if the degradation time is slow, the scale of failure caused by this node is large, and it has a great impact on other nodes.

5.3.2 *Edge Maintenance*

The maintenance behaviour on edges is analysed based on the connection between nodes. Multiple paths always traverse each edge, and different paths are composed of different numbers of edges. In different paths, the number of sequences of the same segment edge in each path is not the same. A higher-level value indicates the edge is located downstream in the path; otherwise, it is located upstream. A network can also be seen as a combination of nodes composed of edges in different hierarchical relationships. The hierarchical value of an edge in the network is the average of the hierarchical values of the same edge in different paths. The average value is equivalent to the value of the edge in the network. This feature of edge layering is referred to as the hierarchical value of edges, which can be expressed by $N_e^{ems} = \sum_{\lambda=1}^{n_e} \frac{h_\lambda}{H_\lambda}$. If the node pair the edge connects fails, it causes the edge to fail. In turn, an edge failure causes the edge's hierarchical value to change. Eventually, increased load transfer pressure causes a larger scale change in the value of the edge. In addition, different load transmission times in all such edges cause the edges to face different pressures from environmental changes. Edge maintenance behaviour is affected by the edge's hierarchical value. The edge maintenance hierarchy can be judged by the change in hierarchical value. Equation (5.25) describes the edge maintenance hierarchy.

$$
\begin{aligned}
N^{ems_f^e} &= \frac{\frac{1}{n_\lambda^e}\sum_{\lambda=1, f\in E}^{n_\lambda^e}\left(\frac{h^e_{f\lambda}}{H^e_{f\lambda}}\right) - \frac{1}{n_\lambda^0}\sum_{\lambda=1, f\in E}^{n_\lambda^0}\left(\frac{h^0_{f\lambda}}{H^0_{f\lambda}}\right)}{T^j} \\
&= \frac{\overline{\left(\frac{h_f^e}{H_f^e}\right)} - \overline{\left(\frac{h_f^0}{H_f^0}\right)}}{T^e} \\
&= \frac{\overline{N^{er_f^e}} - \overline{N^{er_f^0}}}{T^e}
\end{aligned}
$$

$$= \frac{N^{\Delta\overline{er_f^e}}}{T^e}. \tag{5.25}$$

In Eq. (5.25), E represents the set of network edges. $N^{ems_f^e}$ represents the maintenance policy of edge f when any edge e fails. n_λ^e and n_λ^0 represent the total number of paths taken by the edge e after and before the cascade failure, respectively. $h_{f\lambda}^e$ and $H_{f\lambda}^e$ represent the sequence number of edge e on the λ-th path after the cascading failure and the total number of edges included on the λ-th path, respectively. $\frac{h_{f\lambda}^e}{H_{f\lambda}^e}$ represents the hierarchy value of the edge f on the λ-th path that is traversed. $h_{f\lambda}^0$, $H_{f\lambda}^0$ and $\frac{h_{f\lambda}^0}{H_{f\lambda}^0}$ represent the corresponding sequence value, the total number of edges and the hierarchy value on the path before the cascade failure, respectively.

5.3.3 Cooperative Maintenance of Nodes and Edges

Analysis of cooperative maintenance behaviour on nodes and edges starts with the analysis of maintenance on nodes and edges, which affects the overall maintenance of the network. The model for measuring the coordinated maintenance on nodes and edges is shown in Eq. (5.26)

$$\begin{aligned} N^{\text{otms}} &= \chi_n N^{\text{tnms}} + \chi_e N^{\text{tems}} \\ N^{\text{tnms}} &= \sum_{i=1}^{N} \chi_{ni} X(i) \\ N^{\text{tems}} &= \sum_{f=1}^{M} \chi_{ef} X(f). \end{aligned} \tag{5.26}$$

N^{otms} stands for the overall maintenance strategy of the network, N^{tnms} for the maintenance strategy of nodes, and N^{tems} for the maintenance strategy of edges. $\chi_n = \frac{N^{\text{tnms}}}{N^{\text{tnms}}+N^{\text{tems}}}$ is the ratio of N^{tnms} over the sum of N^{tnms} and N^{tems}. $\chi_e = \frac{N^{\text{tnms}}}{N^{\text{tnms}}+N^{\text{tems}}}$, $\chi_{ni} = \frac{N_i^{\text{nms}}}{\sum_{i=1}^{N} \chi_{ni} X(i)}$. The sum of χ_n and χ_e is one, which refers to the ratio of N^{tnms} to the sum of N^{tnms} and N^{tems}, and the ratio of N^{tems} to the sum of N^{tems} and N^{tnms}, respectively. $X(i)$ is determined according to the range set by the following equation: $NC_i = \sum_{q=0}^{k_i^{in}} \left(\frac{Q_i^0 + k_i T^i}{Q_{p_q}^0 - k_{p_q} T^i} \right) - \sum_{q=0}^{k_i^{out}} \left(\frac{Q_{r_q}^0 + k_{r_q} T^i}{Q_i^0 - k_i T^i} \right)$. $X(f)$ is determined according to the range set by the equation: $N_e^{ems} = \sum_{\lambda=1}^{n_e} \frac{h_\lambda}{H_\lambda}$. The range of $X(f)$ is [0,1].

5.4 Summary

In the failure analysis of monolayer networks, we analysed the failures of monolayer network components, including node failures, edge failures and network failures. Furthermore, the cascading failure process of monolayer networks was given, including attack patterns, network failure indicators, load distribution policies, and network failure indicators. In the failure analysis of multilayer networks, we focused on general networks and defined some key metrics that affect the cascading failure process from the perspective of cluster aggregation. Based on this, a cascading failure model based on cluster aggregation was constructed. A quantitative approach was then given to describe the key metrics affecting the failure process. In the maintenance decision of the network, we considered the collaborative maintenance policy of the node edges. The maintenance priority of nodes was also considered to initiate the maintenance of the network.

References

1. Booker G, Torres J, Guikema S, Sprintson A, Brumbelow K (2010) Estimating cellular network performance during hurricanes. Reliab Eng Syst Saf 95(4):337–344
2. Chen L, Dui H, Zhang C (2020) A resilience measure for supply chain systems considering the interruption with the cyber-physical systems. Reliab Eng Syst Saf 199:106869
3. Dui H, Meng X, Xiao H, Guo J (2020) Analysis of the cascading failure for scale-free networks based on a multi-strategy evolutionary game. Reliab Eng Syst Saf 199:106919
4. Estrada E (2012) The structure of complex networks: theory and applications. Oxford University Press, USA
5. Levitin G, Dai YS (2007) Service reliability and performance in grid system with star topology. Reliab Eng Syst Saf 92(1):40–46
6. Levitin G, Xing L (2010) Reliability and performance of multi-state systems with propagated failures having selective effect. Reliab Eng Syst Saf 95(6):655–661
7. Li K, He Y (2017) The complex network reliability and influential nodes. In: AIP conference proceedings, vol 1864. AIP Publishing
8. Ma L, Christou V, Bocchini P (2022) Framework for probabilistic simulation of power transmission network performance under hurricanes. Reliab Eng Syst Saf 217:108072
9. Motter AE, Lai YC (2002) Cascade-based attacks on complex networks. Phys Rev E 66(6):065102
10. Niu YF (2021) Performance measure of a multi-state flow network under reliability and maintenance cost considerations. Reliab Eng Syst Saf 215:107822
11. Rodriguez-Mendez V, Ser-Giacomi E, Hernandez-Garcia E (2017) Clustering coefficient and periodic orbits in flow networks. Chaos: Interdiscip J Nonlinear Sci 27(3):035803
12. Wang Y, Xiao R (2016) An ant colony based resilience approach to cascading failures in cluster supply network. Phys A: Stat Mech Appl 462:150–166
13. Wang Z, Jia L, Ma X, Sun X, Tang Q, Qian S (2022) Accessibility-oriented performance evaluation of high-speed railways using a three-layer network model. Reliab Eng Syst Saf 222:108411
14. Wu J, Baker JW (2020) Statistical learning techniques for the estimation of lifeline network performance and retrofit selection. Reliab Eng Syst Saf 200:106921
15. Wu X, Dong H, Tse CK, Ho IW, Lau FC (2018) Analysis of metro network performance from a complex network perspective. Phys A: Stat Mech Appl 492:553–563

16. Xing L, Dugan JB (2002) Analysis of generalized phased-mission system reliability, performance, and sensitivity. IEEE Trans Reliab 51(2):199–211
17. Zhang L, Wen H, Lu J, Lei D, Li S, Ukkusuri SV (2022) Exploring cascading reliability of multimodal public transit network based on complex networks. Reliab Eng Syst Saf 221:108367
18. Zhou J, Huang N, Coit DW, Felder FA (2018) Combined effects of load dynamics and dependence clusters on cascading failures in network systems. Reliab Eng Syst Saf 170:116–126

Chapter 6
Importance Measures for Resilience Management

Abstract Resilience management for engineered systems has become a topical research field. Resilience management is concerned with readiness and response: the preparedness for negative incidences that may occur and the policies for restoring the failures to the normal state. This chapter aims to investigate the applications of importance measures in resilience management.

Keywords Resilience management · Risk analysis · Network

In recent years, as a manifestation of many modern complex systems, more attention has been increasingly paid to complex networks in science and engineering research [5, 8]. At the same time, natural environment constantly changes and various natural disasters occur frequently. In July 2021, heavy rains in Zhengzhou, China, caused a shut-down of transportation and power systems and incurred immense damage to the city. In early February 2021, snowstorms knocked out the power grid in Texas, leaving more than 4 million customers without water and power supply. The earthquake and tsunami in northeastern Japan in 2011 destroyed some coastal ports and caused significant economic losses. These cases are just the tip of the iceberg. Most large engineered systems present complex characteristics in structure.

There is no doubt that resilient systems can perform better in the face of perturbations. Exploration of methods to improve the resilience for complex networks is obviously an effective means to reduce various losses, which has been studied by many researchers. Hosseini et al. [7] presented a review of recent studies related to defining and quantifying resilience in various disciplines, with a focus on engineered systems. Henry and Ramirez-Marquez [6] described the metrics of network and system resilience, time for resilience and total cost of resilience. They also describe the key parameters necessary to analyse system resilience including disruptive events, component restoration and overall resilience strategy. These metrics serve a similar role as the importance measures in reliability literature. Cerqueti et al. [3] proposed a new measure of network resilience based on the study of the shocks propagation along with the patterns of connections among nodes and the measure being tested on the real-world cases of two important airport systems in the US air traffic network.

Resilience importance measures can quantify the impact of components on system resilience, which extends the applications of reliability importance measures. Both

© The Author(s), under exclusive license to Springer Nature Switzerland AG 2024

H. Dui and S. Wu, *Importance-Informed Reliability Engineering*, Springer Series in Reliability Engineering, https://doi.org/10.1007/978-3-031-52455-4_6

Whitson and Ramirez-Marquez [9] and Barker et al. [1] discussed the relationship between network resilience and ways of exploration of the component importance measure. Baroud et al. [2] demonstrated a time-dependent paradigm for resilience and associated stochastic metrics in a waterway transportation context. Fang et al. [4] proposed two measures, the optimal repair time and the resilience reduction worth, to measure the criticality of the components of a network system from the perspective of their contribution to system resilience.

6.1 A Resilience Measure by Node and Edge Indicators for Monolayer Networks

6.1.1 The Node Resilience for Monolayer Networks

The resilience of a node in a network can be measured by its stability in the face of uncertain attacks or disturbances. The stability can be described by some dynamic metrics, such as the real-time load of the node. Starting from the absolute nature of nodes themselves and the relative nature among nodes, we propose several measures, including the absolute real-time load of nodes, the absolute load transmission rate of nodes, the relative real-time load of nodes, and the relative load transmission rate of nodes. The resilience of the node can then be assessed by constructing and iterating the node resilience matrix.

6.1.2 The Absolute Real-Time Load Transfer Rate

In general, the load flow state of a given node in a network is determined by three time-varying characteristic factors: the time-varying characteristics of the internal load of the node, the time-varying characteristics of the upstream load external to the node, and the time-varying characteristics of the downstream load external to the node. In this section, we can derive the absolute real-time load of a node by first considering the initial internal load and ignoring its topology. Then, by deriving the absolute real-time load of the node, the absolute load transfer rate of the node, i.e., the amount of load transfer of the node per unit time, is obtained. In order to reflect the influence of the upstream and downstream loads of a node on the load flow state, the absolute load transfer rate of a node is denoted by the LIN defined in Sect. 5.2.2 of Chap. 5, and basic form of the absolute load transfer rate as shown in equation, as shown in Eq. (6.1).

$$Q_i^{'}(t) = \begin{cases} \frac{k_{i,\text{out}}}{k_{i,\text{in}}}, & \text{if } k_{i,\text{out}} < k_{i,.\text{in}} \\ 0, & \text{if } k_{i,\text{out}} = k_{i,\text{in}} \\ -\frac{k_{i,\text{out}}}{k_{i,\text{in}}}, & \text{if } k_{i,\text{out}} > k_{i,\text{in}} \end{cases}, \tag{6.1}$$

where $k_{i,\text{out}}$ is the inward value and $k_{i,\text{out}}$ is the outward value. Among them, it is worth noting that in this section, an outward value refers to the number of directed edges from a node, while an inward value refers to the number of directed edges toward a node. Q_i^0 is the initial load of node i, and $k_i = \frac{k_{i,\text{out}}}{k_{i,\text{in}}}$. That is, the absolute real-time load transfer rate is determined by the ratio of the outward value to the inward value. If a node has a high $Q_i^{'}(t)$, the transfer load flows slowly and the outflow flows rapidly, resulting in some extra load that cannot be offset through the process of inflow and outflow transfer load. Therefore, it is reasonable to conclude that the flow rate of the said node is relatively large. Then there are three cases:

- If $\frac{k_{i,\text{out}}}{k_{i,\text{in}}}$ is larger than 1, node i is considered as an upstream node.
- If $\frac{k_{i,\text{out}}}{k_{i,\text{in}}}$ is 1, node i is in the dynamic equilibrium state, and the absolute real time load of the node i is Q_i^0.
- If $\frac{k_{i,\text{out}}}{k_{i,\text{in}}}$ is smaller than 1, node i is considered a downstream node.

Such a metric, which is referred to as the absolute real time load $Q_i(t)$ of the node i, can be given by Eq. (6.2).

$$Q_i(t) = \begin{cases} Q_i^0 + \frac{k_{i,\text{out}}}{k_{i,\text{in}}} t, & \text{if } k_{i,\text{out}} < k_{i,\text{in}} \\ Q_i^0, & \text{if } k_{i,\text{out}} = k_{i,\text{in}}. \\ Q_i^0 - \frac{k_{i,\text{out}}}{k_{i,\text{in}}} t, & \text{if } k_{i,\text{out}} > k_{i,\text{in}} \end{cases} \tag{6.2}$$

6.1.3 The Relative Real-Time Load Transfer Rate

In this section, we propose two new indexes, which are related to topology characteristics, and consider the above three types of time-varying characteristics simultaneously.

Let us consider the ratio of the absolute real time load of one downstream node to the other corresponding upstream node as the relative real time load of the upstream node, which is the amount of load transferred through the downstream node when a unit load is transferred through the corresponding upstream node in a unit of time, regardless of the original load of the two nodes. Consider a node, node i for example, as the object of study. The upstream node is p_i and the downstream node is r_i. Thus, the relative real time load of the node i can be expressed by the following equation.

$$N_i^{rq}(T^i) = Q_i^0 + \sum_{i=0}^{k_{i,\text{in}}} \int_0^{T^i} \frac{Q_i(t)}{Q_{p_i}(t)} dt - \sum_{i=0}^{k_{i,\text{out}}} \int_0^{T^i} \frac{Q_{r_i}(t)}{Q_i(t)} dt. \tag{6.3}$$

In Eq. (6.3), $k_{i,\text{out}}$ and $k_{i,\text{in}}$ denote the values of the outward and inward degrees of node i, respectively. They both are nonnegative integers. For example, in a simple social network, users can follow each other. In this network, each user is a node and the relationship of following each other is a directed edge. If user A follows user B and user C, but no one else follows user A. Then user A's $k_{i,\text{out}}$ is 2 (he follows two people). User A's $k_{i,\text{in}}$ is 0 (no one follows him). $N_i^{rq}(T^i)$ represents the relative real time load of the node i. $N_i^{rq}(T^i)$ is a class of variable upper limit integral functions, and T^i is the period from the beginning of the damage to the end of the damage. Node i may have multiple upstream nodes and downstream nodes, which are referred to as the upstream node set and downstream node set, respectively, and are described by p_i and r_i. $Q_{p_i}(t)$ and $Q_{r_i}(t)$ are the relative real-time load of any node belonging to the upstream node set and the downstream node set, respectively. For the upstream node set, $\int_0^{T^i} \frac{Q_i(t)}{Q_{p_i}(t)}dt$ is used to describe the relative real-time load of node p_i in period T^i, and $\sum_{i=0}^{k_{i,\text{in}}} \int_0^{T^i} \frac{Q_i(t)}{Q_{p_i}(t)}dt$ is a cumulative result, which implies that the total relative real-time load comes from all upstream nodes. This metric can be obtained by deriving the relative real-time load function $N_i^{rq}(T^i)$, and the final result is shown in Eq. (6.4).

$$
\begin{aligned}
N_i^{rq'}(T^i) &= \sum_{i=0}^{k_{i,\text{in}}} \frac{Q_i(T^i)}{Q_{p_i}(T^i)} - \sum_{i=0}^{k_{i,\text{out}}} \frac{Q_{r_i}(T^i)}{Q_i(T^i)} \\
&= \sum_{i=0}^{k_{i,\text{in}}} \frac{Q_i^0 + k_i T^i}{Q_{p_i}^0 - k_{p_i} T^i} - \sum_{i=0}^{k_{i,\text{out}}} \frac{Q_{r_i}^0 + k_{r_i} T^i}{Q_i^0 - k_i T^i},
\end{aligned} \tag{6.4}
$$

where $N_i^{rq'}(T^i)$ is the relative real-time load transfer rate of node i, and the essence of $N_i^{rq'}(T^i)$ is the process of taking the derivative of $N_i^{rq}(T^i)$. For the change of the relative real-time load transfer rate, this metric reflects the rate of change of a node relative to the real-time load transfer rate. Specifically, if the rate of change is slow, there is a little change in the relative real-time load transfer rate. This reflects the node's weak ability to cope with drastic fluctuations while maintaining its own stability. Conversely, this ability can be enhanced if the transfer rate changes more rapidly. The rate of change that affects the interval of change in the transfer rate of the affected node is called the *resilience coefficient* of node i when node i fails. In addition, the rate of change that affects the tolerable interval of the transfer rate of the corresponding node is called the *recovery coefficient* of node i in case of failure of node i. Finally, Eq. (6.5) is obtained after deriving the relative real-time load transfer rate once again.

$$
N_i^{rq''}(T^i) = \sum_{i=0}^{k_{i,\text{in}}} \frac{k_i * Q_{p_i}^0 + k_{p_i} * Q_i^0}{(Q_{p_i}^0 - k_{p_i} * T^i)^2} - \sum_{i=0}^{k_{i,\text{out}}} \frac{k_i * Q_{r_i}^0 + k_{r_i} * Q_i^i}{(Q_i^0 - k_i * T^i)^2}. \tag{6.5}
$$

From Eq. (6.5), it is clear that the relative real-time transmission rate changes constantly over time. When a pair of nodes is identified, the rate of change is only related to the initial failure node and the time span of the cascade failure. The description of nodal resilience will be shown in the next section.

6.1.4 Node Resilience

This section will define node resilience based on the above metrics and give different node resilience matrices considering several cases of initially failed nodes.

Taking node i as an example, when node i fails, it is proposed to measure the resilience of node i by the amount of change in the relative rate of real-time load transfer (QRNs) of nodes within T_i^i. Considering that the resilience coefficients are time-varying, the QRNs can be obtained by integrating the resilience coefficients as shown in Eq. (6.6).

$$N_i^{qrn}(T^i) = \int_{T_{i-1}^i}^{T_{i-1}^i+T_i^i} RQ_i''(T^i)d(T^i) \tag{6.6}$$

In Eq. (6.6), T_i^i is the period of this failure completely through node i. We gave use Eq. (6.6) to calculate the resilience of a certain node. This is not applicable for the case where different nodes are considered as initial failure nodes. Therefore, we start to analyse the node resilience of different nodes. Finally, the matrix of node resilience (MNR) is generated as follows.

$$N^{\text{MNR}} = \begin{bmatrix} 0 & N^{qrn}{}_1^2 & \cdots & N^{qrn}{}_1^i & \cdots & N^{qrn}{}_1^n \\ N^{qrn}{}_2^1 & 0 & \cdots & N^{qrn}{}_2^i & \cdots & N^{qrn}{}_2^n \\ \vdots & \vdots & \ddots & \vdots & \vdots & \vdots \\ N^{qrn}{}_i^1 & N^{qrn}{}_i^2 & \cdots & 0 & \cdots & N^{qrn}{}_i^n \\ \vdots & \vdots & \vdots & \vdots & \vdots & \vdots \\ N^{qrn}{}_n^1 & N^{qrn}{}_n^2 & \cdots & N^{qrn}{}_n^i & \cdots & 0 \end{bmatrix} \tag{6.7}$$

In the above matrix, different rows represent different kinds of initial failure nodes, and each column represents a node number, which contains n nodes in total. On the one hand, a $n \times n$ matrix can be decomposed into an $i \times n$ matrix. Take the first column of the matrix for example, which lists the node resilience of node 1 when all nodes except itself are subsequently used as initial failure nodes. We define the variance of all QRNs in the first column as the nodal resilience of node 1. By analogy, we can obtain the node resilience of all nodes considering all failure types. On the other hand, the $n \times n$ matrix can be decomposed into $n \times 1$ matrices. If we focus only on the first row, then all the row values, i.e., QRNs, reflect the resilience of the remaining nodes in the face of a particular fault caused by node 1. We find that the original value depends not only on the time of load transmission at each node, but also on its resilience factor, which is determined by its topological environment. Secondly, for the node in the relatively central position, it can be expected that the total time of failover is greater than that of other nodes because the node is passed more often. Moreover, the node has a high probability of bearing a huge load variation.

According to the characteristics of clustering theory, multiple nodes contract and form a cluster with each other through a contraction process (or iterative process). Therefore, the overall topology of the network changes during cascading failures. To identify the matrices at different stages, we refer to the matrices mentioned in the previous section as level 0 matrices, so the level 1 matrix (consisting of several level 1 clusters) is given after the first iteration as follows.

$$\begin{bmatrix} 0 & \frac{\sum_{i\in n_2^{(1)}} N^{qrn}{}_i^1}{n_2^{(1)}} & \cdots \frac{\sum_{i\in n_j^{(1)}} N^{qrn}{}_i^1}{n_j^{(1)}} & \cdots \frac{\sum_{i\in n_m^{(1)}} N^{qrn}{}_i^1}{n_m^{(1)}} \\ \frac{\sum_{i\in n_1^{(1)}} N^{qrn}{}_i^2}{n_1^{(1)}} & 0 & \cdots \frac{\sum_{i\in n_j^{(1)}} N^{qrn}{}_i^2}{n_j^{(1)}} & \cdots \frac{\sum_{i\in n_m^{(1)}} N^{qrn}{}_i^2}{n_m^{(1)}} \\ \vdots & \vdots & \ddots & \vdots \\ \frac{1}{n_1^{(1)}} \sum_{i\in n_1^{(1)}} N^{qrn}{}_i^j & \frac{\sum_{i\in n_2^{(1)}} N^{qrn}{}_i^j}{n_2^{(1)}} & \cdots 0 & \cdots \frac{\sum_{i\in n_m^{(1)}} N^{qrn}{}_i^j}{n_m^{(1)}} \\ \vdots & \vdots & \vdots & \ddots \\ \frac{\sum_{i\in n_1^{(1)}} N^{qrn}{}_i^n}{n_1^{(1)}} & \frac{\sum_{i\in n_2^{(1)}} N^{qrn}{}_i^n}{n_2^{(1)}} & \cdots \frac{\sum_{i\in n_j^{(1)}} N^{qrn}{}_i^n}{n_j^{(1)}} & \cdots 0 \end{bmatrix} \quad (6.8)$$

The resilience of a given cluster can be measured by calculating the average value of the node members in the cluster. In the level 1 matrix (a $n \times m$ matrix), m represents the total number of all level 1 matrices. $\frac{1}{n_j^{(1)}} \sum_{i\in n_j^{(1)}} N^{qrn}{}_i^m$, $n_j^{(1)}$ is the number of nodes divided into the j-th cluster of all level 1 clusters. $N^{qrn}{}_i^m$ is the resilience of the i-th cluster when the mth node fails and i should traverse all node members in the cluster. After MNR has iterated for I times, an $n \times 1$ matrix is formed. The transpose of the matrix is shown in Eq. (6.9).

$$\begin{aligned} N^{\mathrm{MNR}^{\mathrm{T}}} = \Big[& \frac{1}{n^{(I)}} \sum_{i\in n^{(I)}} N^{qrn}{}_i^1, \frac{1}{n^{(I)}} \sum_{i\in n^{(I)}} N^{qrn}{}_i^2, \ldots, \\ & \frac{1}{n^{(I)}} \sum_{i\in n^{(I)}} N^{qrn}{}_i^j, \ldots, \frac{1}{n^{(I)}} \sum_{i\in n^{(I)}} N^{qrn}{}_i^n \Big]. \end{aligned} \quad (6.9)$$

For the $n \times 1$ matrix, each row represents all nodes in the network, and the unique column represents the node resilience in the ith-level network, formed after a sufficient number of iterations. Similarly, we take $\frac{1}{n^{(I)}} \sum_{i\in n^{(I)}} N^{qrn}{}_i^j$ as an example, $n^{(I)}$ is the number of nodes belonging to the ith-level cluster, and $N^{qrn}{}_i^j$ represents the node resilience of node i at the initial failure of j. Finally, $\frac{1}{n^{(I)}} \sum_{i\in n^{(I)}} N^{qrn}{}_i^j$ denotes the resilience of the ith-level cluster at the initial failure of node j. Compared with the level 0 matrix, the elements in the ith-level matrix have the same meaning from the perspective of row and column analysis.

6.1.5 Edge Resilience for Monolayer Networks

This section introduces the definition of hierarchical edges, starting from the analysis of edges and edge-related indexes. Iterative treatment of MERs based on clustering theory is then performed. Most importantly, we aim to explore some useful and common laws related to failures.

6.1.5.1 Level Values of Edges

Each edge is always traversed by multiple paths, including several pairs of nodes. Different paths consist of different numbers of edges, and a particular edge belongs to different paths in different orders. We consider the order number of an edge covered by a particular path as the level value of that edge. The higher the level value, the lower the position of the edge. Further, to calculate the level value of an edge connected to the whole network, the level values on each path need to be averaged. The phenomenon of associating the same edge to different sequences is called a hierarchical edge and is represented by Eq. (6.10).

$$q_j = \sum_{i=1}^{n_i} \frac{h_{ij}}{H_{ij}}. \tag{6.10}$$

In Eq. (6.10), q_j denotes the rank value of edge j in the whole network, n_i is the total number of all related paths, H_{ij} is the number of all edges in the j-th path, and h_{ij} is the sequence of the j-th edge dependent on the i-th path. The ratio of the two is the rank value of edge j in the network.

The rate of change of the level value is the relative rate of change of the edge location at the time of failure.

First of all, the failure causes a topological change in the network, which in turn leads to a change in the level value. If the failure is more severe, it will lead to more edge failures on the one hand, which may eventually cause drastic changes. On the other hand, the change in the topological environment around the edge makes the edge more susceptible to greater load transfer pressure. Secondly, the uniqueness of the transfer time for different edges also causes more pressure on the affected edges. Finally, stable edges with a larger amount of variation are considered to have greater resilience. On the contrary, if edges fail because of small variations, they are always inelastic. Therefore, it is reasonable to measure the edge resilience at failure by measuring the rate of change of the horizontal value. The following quantity ${N^{re}}_i^j$ is used to describe such a metric.

$$\begin{aligned} N^{re}{}_i^j &= \frac{\frac{1}{n_\lambda^j}\sum_{\lambda=1,i\in E}^{n_\lambda^j}\frac{h_{i\lambda}^j}{H_{i\lambda}^j} - \frac{1}{n_\lambda^0}\sum_{\lambda=1,i\in E}^{n_\lambda^0}\frac{h_{i\lambda}^0}{H_{i\lambda}^0}}{T^j} \\ &= \frac{\overline{(\frac{h_i^j}{H_i^j})} - \overline{(\frac{h_i^0}{H_i^0})}}{T^j} \\ &= \frac{\overline{q_i^j} - \overline{q_i^0}}{T^j} \\ &= \frac{\overline{\Delta q_i^j}}{T^j} \end{aligned} \tag{6.11}$$

In Eq. (6.11), E denotes the set of edges containing all edges, and $N^{re}{}_i^j$ is the rate of change of the value of edge i when edge j fails, which is the edge resilience of edge i. n_λ^j and n_λ^0 are the total number of paths that edge i can pass through before and after failure, respectively. $h_{i\lambda}^j$ is the sequence of edge i when the load propagates through the λ-th path, and $H_{i\lambda}^j$ is the total number of edges constituted by the λ-th path after failure. Similarly, $h_{i\lambda}^0$, $H_{i\lambda}^0$, and $\frac{h_{i\lambda}^0}{H_{i\lambda}^0}$ have the same meanings and all refer to the period when no failure occurs. $\overline{q_i^j}$ and $\overline{q_i^0}$ denote the horizontal values of edge i before and after a failure, respectively. $N^{re}{}_i^j$ is the amount of change in the level value. It indicates the total time elapsed from the beginning of a failure when edge j is the initial occurrence point of failure. $\overline{\Delta q_i^j}$ is used to quantify the amount of level change, and T^j is the time span during the failure.

6.1.5.2 Edge Resilience

Each edge may fail. In order to explore how to measure the resilience of a particular edge when different initial failure nodes fail, this section proposes definitions of the edge resilience and the edge resilience matrix, respectively. Essentially, the edge resilience is numerically equal to the rate of change of the rank value of a certain node. Furthermore, if we consider each initial failure node as a case, MER (the matrix of edge resilience) can be formed as shown in Eq. (6.12).

$$\mathbf{N}^{MER} = \begin{bmatrix} 0 & N^{re}{}_1^2 & \cdots & N^{re}{}_1^i & \cdots & N^{re}{}_1^e \\ N^{re}{}_2^1 & 0 & \cdots & N^{re}{}_2^i & \cdots & N^{re}{}_2^e \\ \vdots & \vdots & \ddots & \vdots & \vdots & \vdots \\ N^{re}{}_i^1 & N^{re}{}_i^2 & \cdots & 0 & \cdots & N^{re}{}_i^e \\ \vdots & \vdots & \vdots & \vdots & \ddots & \vdots \\ N^{re}{}_e^2 & N^{re}{}_2^e & \cdots & N^{re}{}_2^i & \cdots & 0 \end{bmatrix} \tag{6.12}$$

In a certain MER, each row represents a failure case, the REs of each remaining edge is shown, and each column gives the REs of the same edge considering all failure cases.

On the one hand, the matrix of order $e \times e$ can be decomposed into a matrix of order $1 \times e$ rows. As an example, the first column of the matrix lists the RE of edge 1 considering e. Overall, it is appropriate to define the deviation of all REs belonging to column 1 as the final resilience of edge 1. By analogy, the final resilience of others in the network can be obtained.

On the other hand, the $e \times e$ order matrix ($e \times e$ order means that the matrix has e rows and e columns) can also be reduced to an $e \times 1$ order column matrix. For example, if we focus on the first row, we can obtain the REs of all remaining edges when edge 1 initially fails in some failure. It can be seen that the REs of different edges depends only on the number of traversable paths. In general, midstream edges have an important role and are more likely to be connected by other edges, so the rate of change of the level values will be larger with higher probability.

The process is similar to the iteration of the node elasticity matrix. The formatting of the clusters is also a process of reorganisation of the edge structure. For the sake of distinction, we name the initial matrix as a 0-level matrix. After that, the first shrinkage aggregation of edges is performed and Eq. (6.13) shows a new 1-level matrix.

$$
\begin{bmatrix}
0 & \frac{1}{n_2^{(1)}}\sum_{i\in n_2^{(1)}} N^{re1}_i & \cdots \frac{1}{n_j^{(1)}}\sum_{i\in n_j^{(1)}} N^{re1}_i & \cdots \frac{1}{n_m^{(1)}}\sum_{i\in n_m^{(1)}} N^{re1}_i \\
\frac{1}{n_1^{(1)}}\sum_{i\in n_1^{(1)}} N^{re2}_i & 0 & \cdots \frac{1}{n_j^{(1)}}\sum_{i\in n_j^{(1)}} N^{re2}_i & \cdots \frac{1}{n_m^{(1)}}\sum_{i\in n_m^{(1)}} N^{re2}_i \\
\vdots & \vdots & \ddots & \vdots \\
\frac{1}{n_1^{(1)}}\sum_{i\in n_1^{(1)}} N^{rej}_i & \frac{1}{n_2^{(1)}}\sum_{i\in n_2^{(1)}} N^{rej}_i & \cdots 0 & \cdots \frac{1}{n_m^{(1)}}\sum_{i\in n_m^{(1)}} N^{rej}_i \\
\vdots & \vdots & \vdots & \ddots \\
\frac{1}{n_1^{(1)}}\sum_{i\in n_1^{(1)}} N^{ree}_i & \frac{1}{n_2^{(1)}}\sum_{i\in n_2^{(1)}} N^{ree}_i & \cdots \frac{1}{n_j^{(1)}}\sum_{i\in n_j^{(1)}} N^{ree}_i & \cdots 0
\end{bmatrix}
\tag{6.13}
$$

The 1-level matrix we are dealing with is essentially an $e \times m$ matrix, where e is the total number of edges in the network and m is the total number of elementary element 1-level clusters. Taking $\frac{1}{n_2^{(1)}}\sum_{i\in n_2^{(1)}} N^{rej}_i$ as an example, $n_2^{(1)}$ is numerically equal to the number of 1-level clusters in the network after the 1st iteration, and RE_i^j is used to describe the elasticity of the ith edge when edge j is used as the initial failed edge. Similar to the initial 0-level matrix, the variance of each column in the 1-level matrix represents the elasticity of the corresponding 1-level cluster, while the rank of each row reflects the edge importance of all the remaining edges at the initial failure of an edge. It is worth noting that we always have different rankings due to the different cases of initial failure edges, which need to be combined.

After the first iteration, we keep aggregating the latest matrix, and finally obtain the transpose of the matrix as follows.

$$N^{MER^T} = [\frac{1}{n^{(I)}} \sum_{i \in n^{(I)}} N^{re1}_{i}, \frac{1}{n^{(I)}} \sum_{i \in n^{(I)}} N^{re2}_{i}, \ldots, \frac{1}{n^{(I)}} \sum_{i \in n^{(I)}} N^{rej}_{i}, \ldots, \frac{1}{n^{(I)}} \sum_{i \in n^{(I)}} N^{ree}_{i}]. \tag{6.14}$$

In Eq. (6.14), each element denotes one failure considering the initial failure edge, and the unique column denotes the ith-level matrix containing the unique level 1 cluster after the i-th iteration. $n^{(I)}$ is the number of edges in the final cluster, and N^{rej}_{i} has the same meaning.

6.1.5.3 Monolayer Network Resilience

This section investigates methods to integrate the node resilience and the edge resilience and then measure the resilience of an entire network. Moreover, it shows the difference between the resilience of node and that of the edge resilience, and explain how they influence the network in different ways.

Node resilience is described as the amount of variation in the relative rate of real-time load shifting of nodes. In other words, it is a fluctuation. Firstly, node resilience is measured continuously during the failure process, so it can be considered as a process indicator. Secondly, we obtain the edge resilience without considering the situation of upstream or downstream. The indicator exactly reflects the changes of surrounding environment, which is different from the node resilience. Moreover, from the perspective of the level of impact, a failure caused by one node may lead to the failure of all connected edges, while a failure caused by a single edge does not necessarily lead to the failure of all connected nodes, unless the affected node has only one connected edge, i.e., the connected edge that is the initial failure edge. Therefore, node resilience has a greater impact on network stability compared to edge resilience.

The resilience of an edge is essentially the rate of change of the level value after the corresponding edge failure. In what follows, we first present the metrics for two cases, which are the time before and after a cascade failure. The rate of change is not a real-time rate but an average rate, and RE (Resilience of Edge) is a result-oriented metric. We then obtain the edge resilience without considering the upstream or downstream cases. This metric precisely reflects the changes in the surrounding environment, unlike the node resilience. Therefore, it belongs to a kind of absolute metrics. The loss caused by edge failure is smaller than the loss caused by node failure, and the impact of edge resilience on the normal operation of the network is always weak.

In this section, we start with analysing the relationship between the two nodes and then give a measure of network resilience. In view of this, a method that applies the idea of weighting is used to provide a comprehensive measure. Finally, we measure the network resilience by Eq. (6.15).

$$N^{\text{resilience}} = \lambda_n \sum_{i=1}^{N} \lambda_{ni} X(i) + \lambda_e \sum_{j=1}^{M} \lambda_{ej} X(j) \tag{6.15}$$

In Eq. (6.15), $N^{\text{resilience}}$ denotes the network resilience. The sum of λ_n and λ_e is equal to 1. λ_n represents the ratio of $\sum_{i=1}^{N} \lambda_{ni} X(i)$ to the sum of $\sum_{i=1}^{N} \lambda_{ni} X(i)$ and $\sum_{j=1}^{M} \lambda_{ej} X(j)$. λ_e represents the ratio of $\sum_{j=1}^{M} \lambda_{ej} X(j)$ to the sum of $\sum_{i=1}^{N} \lambda_{ni} X(i)$ and $\sum_{j=1}^{M} \lambda_{ej} X(j)$. λ_n is a weight determined based on the QRNs located in the same column in a given MNR, and λ_{ej} is derived in the same way as that for MERs. $X(i)$ is a weight determined by the extreme differences in each row of the MNRs at the time of failure, with each node initially failing. Similarly, $X(j)$ is derived in the same way. In addition, it can be seen that both $\sum_{i=1}^{N} \lambda_{ni} X(i)$ and $\sum_{j=1}^{M} \lambda_{ej} X(j)$ take values in the interval [0, 1] and $N^{\text{resilience}}$ takes value in interval [0, 0.5].

6.2 Residual Resilience Assessment for Monolayer Infrastructure Networks

6.2.1 Definition and Quantification of Resilience of Infrastructure Network

Definition 6.1 Resilience is defined as the ability of a smart grid to resist, adapt and quickly recover to normal and stable operation states after being attacked by disasters.

The lifetime of a disaster management can be divided into several phases, as discussed below.

- Disaster prevention phase: This is when the infrastructure network is functioning normally. During this phase, advanced weather forecasting and decision support systems can be used to prevent and prepare for disasters.
- Disaster occurrence phase: In this phase, we assume that the infrastructure network performance is in a nonlinear degradation state and the degradation rate is gradually accelerated. The disaster occurs in t_1 and the infrastructure network performance is affected to some extent. The degree of impact depends on the severity of the disaster attack and the resistance capability of the infrastructure network.
- Degraded operation phase: We assume that the infrastructure network performance is in a stable degraded state during this phase. After a disaster, the infrastructure network absorbs the disaster attack and operates in a degraded state. Before the failure recovery, the facility managers can help the infrastructure network adapt to the disaster attack through a series of optimisation operations based on the failure state.

- Failed component recovery phase: We assume that the performance of the infrastructure network gradually improves in this phase. As the recovery proceeds, the improvement of infrastructure network performance gradually diminishes. At t_3, the failed component starts to be repaired and the infrastructure network operation state is gradually restored.
- Stable operation phase: the repair of faulty components is completed and the infrastructure network gradually returns to a stable operating state.

Based on Definition 6.1, below presents the concept of residual resilience.

Definition 6.2 Residual resilience is quantified as the difference between the current resilience and the optimal resilience, is defined by

$$\begin{aligned} R'^{\text{resilience}}(t) =& \frac{Q^{\text{loss}}(t) - Q^{\text{recovery}}(t)}{Q^{\text{loss}}(t)} \\ =& \frac{\int_{t_1}^{t} (Q^*(t) - Q(t_3))\, dt - \int_{t_1}^{t} (Q(t) - Q(t_3))\, dt}{\int_{t_1}^{t} (Q^*(t) - Q(t_3))\, dt} \\ =& 1 - \frac{\int_{t_1}^{t} (Q(t) - Q(t_3))\, dt}{\int_{t_1}^{t} (Q^*(t) - Q(t_3))\, dt}, \end{aligned} \tag{6.16}$$

where $Q^{\text{loss}}(t)$ is expressed as the integral of the difference between the optimal performance of the infrastructure network and the worst performance of the infrastructure network during the time period (t_1, t), and $Q^{\text{recovery}}(t)$ is expressed as the integral of the difference between the current performance of the infrastructure network and the worst performance of the infrastructure network.

When $Q(t) = Q(t_3)$, $R'^{\text{resilience}}(t) = 1$, the value of $R'^{\text{resilience}}(t)$ is in the interval $[0,\ 1]$, which means that the infrastructure network is still in the post-disaster state and no failed components are successfully repaired; when $Q(t) = Q^*(t)$, $R'^{\text{resilience}}(t) = 0$, which is the ideal situation when the post-disaster infrastructure network recovers to the target state. The closer the value of $R'^{\text{resilience}}(t)$ is to 0, the better the recovery. This definition quantifies the scale and speed of the recovery of an infrastructure network. Compared with Barker et al. [1], the definition of resilience is memorised because it takes into account the history of infrastructure network recovery performance.

6.2.2 *Residual Resilience Optimisation Model for the Infrastructure Network*

A catastrophic event may cause one or more components in the infrastructure network to fail. When multiple components fail after a catastrophic event, the main challenge is to determine the order in which the failed components should be restored. The goal is to maximise the resilience of the infrastructure network and to restore it to an optimal state within a certain period. This section focuses on the impact of individual components on the remaining resilience of different states of an infrastructure network. Thus, the optimal repair order of a failed component group can result in the smallest residual resilience within the recovery time, thus reducing economic losses.

This section makes the following assumptions:

- The states of the components in a infrastructure network are statistically independent of each other, and only two states exist: normal and failure.
- After a disaster, the failed components in the infrastructure network can be restored within the same time.
- The recovery of a faulty component is a discrete event, i.e., only one faulty component can be repaired in a given time.

The infrastructure network is defined by $\boldsymbol{G(N, L)}$, where $\boldsymbol{N}$ denotes the set of nodes in the infrastructure network, and $\boldsymbol{L}$ denotes the set of edges in the infrastructure network ($\boldsymbol{L} \subset \{(i, j) : i, j \in N, i \neq j\}$). The set of nodes in the infrastructure network includes the set of supply nodes $\boldsymbol{N_S}$, the set of demand nodes $\boldsymbol{N_D}$. $\boldsymbol{C_0^+}$ $\left(\boldsymbol{C_0^+} \in R^+\right)$ denotes the set of capacity of the components in the infrastructure network. The capacities of edge ij, supply node $i \in \boldsymbol{N_S}$ and demand node $j \in \boldsymbol{N_D}$ are denoted by P_{ij}, $P_i^S$and P_j^D, respectively. $\boldsymbol{E}$ is defined as the set of infrastructure network components and $\boldsymbol{E}'$ is modelled as the set of failed components in the infrastructure network.

We now determine the repair order of the set of failed components in a given time period with the objective of minimising the remaining elasticity. Thus, a time set $t \in \{0, 1, 2, 3 \ldots T\}$ $(T \in Z^+)$ consists of multiple discrete time periods, each of which repairs only one failed component.

$Q_j(t)$ is modelled as the demand node j. The aim is to maximise the flow of the demand node.

$$Q(t) = \sum_{j \in N_D} Q_j(t) \tag{6.17}$$

The residual resilience is then obtained by plugging Eq. (6.18) into Eq. (6.17).

$$R'^{\text{resilience}}(t) = \frac{T\left(\sum_{j \in N_D} P_j^D - Q_0\right) - \sum_{t \in T}\left[\sum_{j \in N_D} Q_j(t) - Q_0\right]}{T\left(\sum_{j \in N_D} P_j^D - Q_0\right)} \tag{6.18}$$

where $\sum_{j \in N_D} P_j(t)$ denotes that the demands of all demand nodes in N_D are fully satisfied, i.e., $\sum_{j \in N_D} P_j(t)$. When $t = t_3$, the system starts to repair. $Q(t_3)$ can be expressed as Q_0. Therefore, the optimisation model with the objective of minimising the residual resilience over the recovery time spanning the distance T is as follows.

$$\min R'^{\text{resilience}}(t) = \min \frac{T\left(\sum_{j\in N_D} P_j^D - Q_0\right) - \sum_{t\in T}\left[\sum_{j\in N_D} w_j(t) - Q_0\right]}{T\left(\sum_{j\in N_D} P_j^D - Q_0\right)} \tag{6.19}$$

subject to

$$\sum_{(i,j)\in E} Q_{ij}(t) - \sum_{(j,i)\in E} Q_{ji}(t) \le P_i^S,\ i \in N_S, \forall t \tag{6.20}$$

$$\sum_{(i,j)\in E} Q_{ij}(t) - \sum_{(j,i)\in E} Q_{ji}(t) = Q_j(t),\ j \in N_D, \forall t \tag{6.21}$$

$$0 \le Q_j(t) \le P_j^D, j \in N_D, \forall t \tag{6.22}$$

$$0 \le Q_{ij}(t) \le \mu_{ij}(t) P_{ij}, (i, j) \in E, \forall t \tag{6.23}$$

$$0 \le Q_{ij}(t) \le \mu_i(t) P_{ij}, (i, j) \in E,\ i \in N, \forall t \tag{6.24}$$

$$0 \le Q_{ij}(t) \le \mu_i(t) P_{ij}, (i, j) \in E, j \in N, \forall t \tag{6.25}$$

$$\mu_{ij}(t) - \mu_{ij}(t+1) \le 0, (i, j) \in E, \forall t \tag{6.26}$$

$$\mu_i(t) - \mu_i(t+1) \le 0, i \in E, \forall t \tag{6.27}$$

$$\sum_{i\in E'} [\mu_i(t) - \mu_i(t-1)] + \sum_{(i,j)\in E'} \left[\mu_{ij}(t) - \mu_{ij}(t-1)\right] \le 1, \forall t \tag{6.28}$$

$$\mu_{ij}(t) \in \{0, 1\},\ (i, j) \in E, t \in \forall t \tag{6.29}$$

$$\mu_i(t) \in \{0, 1\},\ i \in E, t \in \forall t \tag{6.30}$$

$$\mu_{ij}(0) = 0,\ (i, j) \in E' \tag{6.31}$$

$$\mu_i(0) = 0,\ i \in E' \tag{6.32}$$

where $Q_{ij}(t)$ denotes the flow from supply node i to demand node j at t. $\mu_{ij}(t)$ and $\mu_i(t)$ denote the states of edge ij and supply node i, respectively, where 1 indicates that the infrastructure network is in a normal state and 0 indicates a failure. Equation (6.20) ensures that the supply demand of $i(i \in \boldsymbol{N_S})$ does not exceed its supply capacity; Eqs. (6.21) and (6.22) indicate that demand node $j(j \in \boldsymbol{N_D})$ does not exceed its demand; Eqs. (6.23)—(6.25) indicate that edge ij and supply node i must not exceed the capacity that can be passed in the current state; Eqs. (6.26) and (6.27) show that the situation of edge ij and supply node i will obtain better with time, i.e., once the fault is repaired no further failure will occur; Eq. (6.28) indicates that only one failed component can be repaired in a given time interval; Eqs. (6.29)–(6.32) are used to ensure that only two states of operation and failure exist for edge ij and supply node i, and all components in E' fail in the initial state.

6.3 Resilience Importance for the Monolayer Network

6.3.1 Performance Change of Monolayer Network

If nodes or routes are damaged, the performance of a monolayer network will be reduced. The performance curve of a monolayer network is shown in Fig. 6.1, where $A(t)$ ($A(t) \in [0, 1]$) is the performance of the monolayer network after being attacked at time t. During the recovery period of the failed node, the performance of the monolayer network gradually recovers.

In Fig. 6.1, the network is in the normal state during (t_0, t_a), and the nodes in the network are in the normal state, the maximum performance of the network is $A_{target}(t)$. At time t_a, the network is attacked, multiple nodes fail, and the network performance rapidly decreases to the lowest state $A(t_b)$. At time t_b, the network begins to be repaired. From t_b to t_γ, the network is in the recovery phase. At time t_γ, the network completes the maintenance work, and the performance of the network after recovery is $A(t_\gamma)$.

The loss performance of the monolayer network is defined as

$$Q^{\text{loss}}(t) = A^{\text{target}}(t) - A(t_b). \tag{6.33}$$

The recovery performance of monolayer network is defined as

$$Q^{\text{recovery}}(t) = A\left(t_\gamma\right) - A(t_b). \tag{6.34}$$

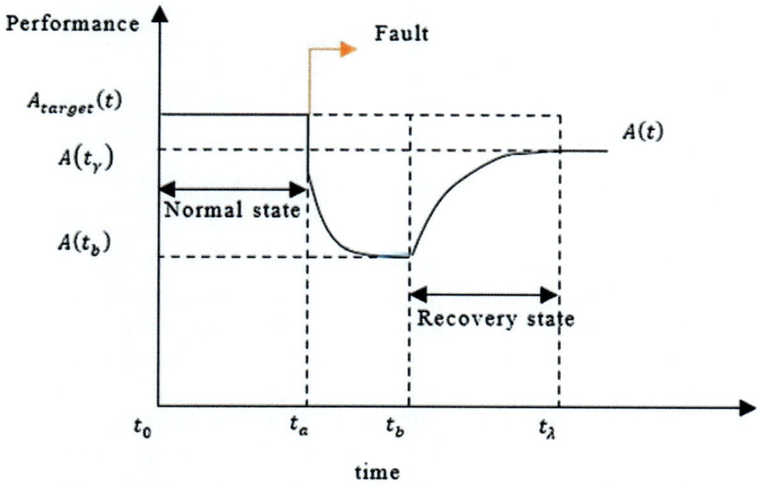

Fig. 6.1 Performance curve of land transport network

Node i has two states with $\tau_i = 1$ indicating that the node i is working and $\tau_i = 0$ indicating that it is in the failed state. The loss performance of the monolayer network when the node i is working can be expressed as

$$Q^{\text{loss}}(t)_{\tau_i=1} = A_{target}(t) - A\left(t_b\right)_{\tau_i=1}, \tag{6.35}$$

where $A\left(t_b\right)_{\tau_i=1}$ indicates the performance of the monolayer network when all the failed nodes are working normally.

The loss performance of the network when node i fails can be expressed as

$$Q^{\text{loss}}(t)_{\tau_i=0} = A_{target}(t) - A(t_b)_{\tau_i=0}, \tag{6.36}$$

where $A(t_b)_{\tau_i=0}$ represents the performance of the monolayer network when other nodes are working normally and only node i fails.

As mentioned above, the recovery performance of the network when node i is working can be expressed as

$$Q^{\text{recovery}}(t)_{\tau_i=1} = A\left(t_\gamma\right)_{\tau_i=1} - A\left(t_b\right), \tag{6.37}$$

where $A\left(t_\gamma\right)_{\tau_i=1}$ represents the performance of the monolayer network when other failed nodes do not return to normal and only node i returns to normal.

The recovery performance of the monolayer network when node i fails can be expressed as

$$Q^{\text{recovery}}(t)_{\tau_i=0} = A\left(t_\gamma\right)_{\tau_i=0} - A\left(t_b\right), \tag{6.38}$$

where $A\left(t_\gamma\right)_{\tau_i=0}$ indicates the performance of the monolayer network when none of the failed nodes has returned to normal operation.

6.3.2 Resilience Importance of Monolayer Network

The influence of different nodes on the performance of a monolayer network is also different. The loss of node performance has a positive proportion to the impact on the vulnerability of the monolayer network. The faster the node performance recovery, the greater the impact of the normal operation of the node on the network performance recovery.

When the nodes in the monolayer network are in different running states, the performance loss of the whole network will change differently. Based on the Birnbaum importance measure, the formula for measuring the loss importance measure of the node i is given by

$$\begin{aligned} I_i^{\text{loss}}(t) &= Q^{\text{loss}}(t)_{\tau_i=0} - Q^{\text{loss}}(t)_{\tau_i=1} \\ &= \left(A^{\text{target}}(t) - A(t_b)_{\tau_i=0}\right) - \left(A^{\text{target}}(t) - A(t_b)_{\tau_i=1}\right) \\ &= A(t_b)_{\tau_i=1} - A(t_b)_{\tau_i=0}. \end{aligned} \tag{6.39}$$

I_i^{loss} indicates the effect of the state change of node i on the performance loss of the monolayer network. By comparing the loss importance measure value of each node, the change of network vulnerability can be evaluated. By calculating the loss importance measure of nodes, the importance of nodes can be predicted, and preventive maintenance of nodes with greater importance can be carried out in advance. $\max\{I_i^{\text{loss}},\ i = 1,\ 2, .., n\}$ indicates the node which has the greatest impact on the vulnerability of the monolayer network.

When the nodes are in different running states, the performance recovery of the whole monolayer network will change differently. Based on the Birnbaum importance measure, the recovery importance measure of node i is given by

$$\begin{aligned} I_i^{\text{recovery}}(t) &= Q^{\text{recovery}}(t)_{\tau_i=1} - Q^{\text{recovery}}(t)_{\tau_i=0} \\ &= \left(A(t_\lambda)_{\tau_i=1} - A(t_b)\right) - \left(A(t_\lambda)_{\tau_i=0} - A(t_b)\right) \\ &= A(t_\lambda)_{\tau_i=1} - A(t_\lambda)_{\tau_i=0}, \end{aligned} \tag{6.40}$$

where $I_i^{\text{recovery}}(t)$ indicates the impact of state changes of node i on the performance recovery of a monolayer network. The change of network recovery performance can be evaluated by comparing the recovery importance values of each node. The value of $\max\{I_i^{\text{recovery}}(t),\ i = 1,\ 2, .., n\}$ represents the node that has the greatest impact on the recoverability of the monolayer network.

Resilience importance measure takes into account the failure process and the recovery process. In a monolayer network, the resilience importance measure of the node i can be expressed as the ratio of the recovery importance measure to the loss importance measure.

$$I_i^{\text{resilience}}(t) = \frac{I_i^{\text{recovery}}(t)}{I_i^{loss}(t)} = \frac{Q^{\text{recovery}}(t)_{\tau_i=1} - Q^{\text{recovery}}(t)_{\tau_i=0}}{Q^{\text{loss}}(t)_{\tau_i=0} - Q^{\text{loss}}(t)_{\tau_i=1}}. \tag{6.41}$$

$I_i^{\text{resilience}}(t)$ refers to the resilience importance measure of each node in a monolayer network, which is used to evaluate the impact of different nodes on the network performance. The value $\max\{I_i^{\text{resilience}}(t),\ i = 1,\ 2, .., n\}$ represents the node that has the greatest impact on the performance of the monolayer network.

References

1. Barker K, Ramirez-Marquez JE, Rocco CM (2013) Resilience-based network component importance measures. Reliab Eng Syst Saf 117:89–97
2. Baroud H, Barker K, Ramirez-Marquez JE et al (2014) Importance measures for inland waterway network resilience. Transp Res Part E: Logis Transp Rev 62:55–67
3. Cerqueti R, Ferraro G, Iovanella A (2019) Measuring network resilience through connection patterns. Reliab Eng Syst Saf 188:320–329
4. Fang YP, Pedroni N, Zio E (2016) Resilience-based component importance measures for critical infrastructure network systems. IEEE Trans Reliab 65(2):502–512
5. Gao J, Barzel B, Barabási AL (2016) Universal resilience patterns in complex networks. Nature 530(7590):307–312
6. Henry D, Ramirez-Marquez JE (2012) Generic metrics and quantitative approaches for system resilience as a function of time. Reliab Eng Syst Saf 99:114–122
7. Hosseini S, Barker K, Ramirez-Marquez JE (2016) A review of definitions and measures of system resilience. Reliab Eng Syst Saf 145:47–61
8. Nikinmaa L, Lindner M, Cantarello E, Jump AS, Seidl R, Winkel G, Muys B (2020) Reviewing the use of resilience concepts in forest sciences. Curr Fores Rep 6:61–80
9. Whitson JC, Ramirez-Marquez JE (2009) Resiliency as a component importance measure in network reliability. Reliab Eng Syst Saf 94(10):1685–1693

Chapter 7
Case Studies

Abstract While the previous five chapters introduced concepts on importance measures and discussed their potential applications in the real world, few case studies have shown how the concepts could be applied in practice. The engineered systems discussed in this chapter include wind power systems, satellite attitude control systems, rocket vertical assembly and test plant systems, multi-level disasters, and land transport networks, which are used for illustrating the concepts introduced in the preceding chapters.

Keywords Markov process · Importance measures · Constraint programming

This chapter applies the proposed methods to practical cases.

7.1 Wind Power Systems

The operation of wind power systems is characterised by randomness, intermittency and volatility, and susceptible to failures due to severe external environmental impact [12]. To reduce the probability of the occurrences of serious economic losses, this section analyses the reliability of wind power systems with the approaches proposed in Chap. 2. Firstly, the Birnbaum importance measure (BIM) [1] and the integrated importance measure (IIM) [29] are used to analyse the influence of the nodes on the reliability of wind power systems, which is optimised by using the importance gradient [29]. The fastest direction of reliability improvement of the wind power system is then assessed.

7.1.1 Reliability of Wind Power Systems

A wind power system is a clean grid system that converts wind energy into electrical energy. For ease of analysis, it is specified that each wind turbine has the same power rating and the remaining nodes are all power demand nodes. Figure 7.1 shows a typical wind power network diagram, consisting of the power supply node (wind turbine), the power demand node and the transmission edge between the nodes.

© The Author(s), under exclusive license to Springer Nature Switzerland AG 2024
H. Dui and S. Wu, *Importance-Informed Reliability Engineering*, Springer Series in Reliability Engineering, https://doi.org/10.1007/978-3-031-52455-4_7

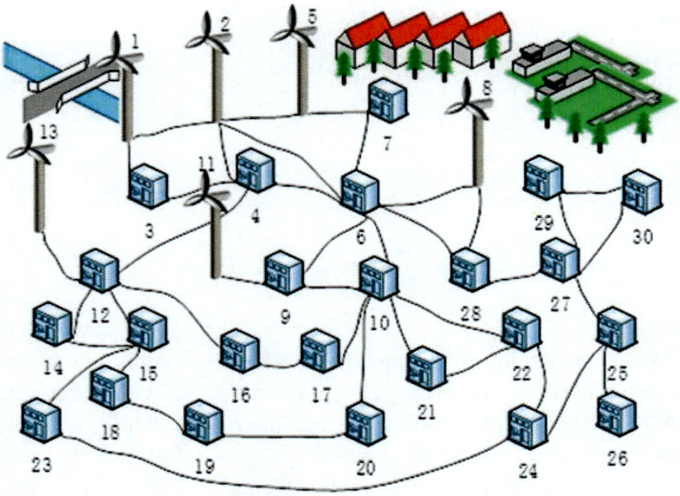

Fig. 7.1 Wind power network diagram

Fig. 7.2 Wind speed versus actual power output

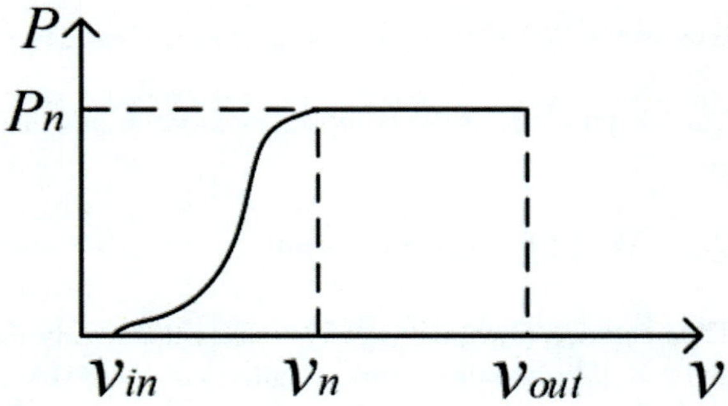

Wind blows the fan blades of wind turbine P_n to rotate and produce electric power. When the wind speed is too high or too low, it may cause the wind turbine to fail and the reliability of the wind power system to decrease. The output power of a wind turbine is therefore closely related to the wind speed. The actual rated power output of a wind turbine can be described in terms of the cut-in wind speed v_{in}, the rated wind speed v_n, and the cut-out wind speed v_{out}. The actual output power P_w of a wind turbine under normal air density conditions against wind speed is shown in Fig. 7.2.

When the wind speed is less than the cut-in wind speed or greater than the cut-out wind speed, the wind turbine will automatically shut down or the tower will collapse or the impeller will fly. Therefore, in both cases the wind turbine is failed and the output power is 0. The actual output power of the wind turbine relates to the wind speed by funcction shown in Eq. (7.1).

$$P_w = \begin{cases} 0, & \text{if } v < v_{in} \text{ or } v > v_{out} \\ P_n\left(\frac{v-v_{in}}{v_n-v_{in}}\right), & \text{if } v_{in} \le v \le v_n \\ P_n, & \text{if } v_n \le v \le v_{out} \end{cases} . \tag{7.1}$$

Unlike the reliability of other engineered systems, the reliability of a wind power system is measured by the level of supply performance rather than its lifetime as the independent variable. In this case, the level of performance depends on the per capita demand for electricity and the amount of electricity generated per unit. The reliability, as defined in Eq. (7.2), of a wind power system is therefore defined as the probability of achieving the required electricity consumption by the customer within a specified period of time under specified operating conditions.

$$\begin{aligned} R(t, d) &= \Pr\{T \le t, Q \ge d\} \\ &= \Pr\left\{\frac{Q}{T} \ge \frac{d}{t}\right\}, \end{aligned} \tag{7.2}$$

where t and d are the electricity demand time and demand power consumption, respectively; the random variables T and Q are the generation time and generation capacity, respectively; Q/T denotes the generation capacity per unit of time and is equivalent to the actual output power P_w and d/t denotes the demand power per unit of time and is denoted by P_d. Then the reliability of the wind power generation system

$$R(P_d) = \Pr\{P_w \ge P_d\}. \tag{7.3}$$

In a wind power series system, the wind power generation system is composed of n wind turbines in series. A wind turbine output power with less than the cut-in wind speed or greater than the cut-out wind speed will cause the equipment to fail, which in turn will cause the wind power system to fail. Assume that the output power of the turbine i is P_{wi} and the reliability is $R_i(P_d), i = 1, 2, \ldots, n$.

In a series system, current flows sequentially from one component to the next, so the power of the current through the entire system is limited by the weakest component through which the current flows. This is because the minimum value of power generated by the component determines the maximum output power of the entire system. From the series connection, the power output of the wind power system is

$$P_w = \min\{P_{w1}, P_{w2}, \ldots, P_{wn}\}. \tag{7.4}$$

The reliability of a series power system is given by

$$\begin{aligned} R(P_d) &= \Pr\{\min\{P_{w1}, P_{w2}, \ldots, P_{wn}\} > P_d\} \\ &= \Pr\{P_{w1} > P_d, P_{w2} > P_d, \ldots, P_{wn} > P_d\} \\ &= \prod_{i=1}^{n} R_i(P_d). \end{aligned} \tag{7.5}$$

In a wind power parallel system, suppose the wind power system composed of n wind turbines is connected in parallel. The failure of this equipment is caused when the output power of all wind turbines is less than the cut-in wind speed or greater than the cut-out wind speed, which in turn causes the wind power system to fail. Assume that the output power of the turbine i is P_{wi} and the reliability is $R_i(P_d), i = 1, 2, \ldots, n$. It can be seen that the wind power system generates:

$$P_w = \max\{P_{w1}, P_{w2}, \ldots, P_{wn}\}. \tag{7.6}$$

The reliability of a parallel power system is:

$$\begin{aligned} R(P_d) &= \Pr\{\max\{P_{w1}, P_{w2}, \ldots, P_{wn}\} > P_d\} \\ &= 1 - \Pr\{\max\{P_{w1}, P_{w2}, \ldots, P_{wn}\} \le P_d\} \\ &= 1 - \Pr\{P_{w1} \le P_d, P_{w2} \le P_d, \ldots, P_{wn} \le P_d\} \\ &= 1 - \prod_{i=1}^{n} [1 - R_i(P_d)]. \end{aligned} \tag{7.7}$$

7.1.2 Importance Measure Gradients for Wind Power Systems

The BIM is applied to analyse the reliability of wind power systems for measuring the extent to which the changes in the state of a node affect the reliability of the system. The BIM of node i of the wind power system is denoted by I_i^B, and the measure of the node lifetime is converted to the demand power consumption and is expressed as

$$\begin{aligned} I_i^B(P_d) &= \frac{\partial R(P_d)}{\partial R_i(P_d)} \\ &= \frac{\partial \Pr\{\Phi(P_w) = 1\}}{\partial \Pr\{P_{wi} = 1\}} \\ &= \Pr\{\Phi(P_w) = 1 | P_{wi} = 1\} - \Pr\{\Phi(P_w) = 1 | P_{wi} = 0\}, \end{aligned} \tag{7.8}$$

where $I_i^B(P_d)$ denotes the impact of a failure of node i on the system reliability, $R(P_d)$ is the wind power system reliability, $R_i(P_d)$ is the reliability of node i, and $\Pr\{\Phi(P_w) = 1\}$ is the probability that the power generated by the wind power system meets the power demanded by the users. $\Pr\{P_{wi} = 1\}$ denotes the probability that the node works normally and $\Pr\{P_{wi} = 0\}$ denotes the probability that the node fails. By comparing the BIM values of each node, the change in the reliability of the wind power system can be assessed. Thus, $\max\left\{I_i^B, i = 1, 2, \ldots, n\right\}$ denotes the node with the greatest impact on the reliability of the wind power system.

The integrated importance measure indicates the degree of impact on system reliability when a component fails in different states, with state failure superposition. The degree of the influence of the component state distribution probability and the component state transfer rate on system reliability is considered. The integrated importance in the wind power system indicates the impact of nodes at different states of failure superposition on the reliability of the wind power system. The IIM of the ith node of the wind power system is denoted by $I_i^{im}(P_d)$, which is given below

$$
\begin{aligned}
I_i^{im}(P_d) &= \sum_{m=1}^{M} I_{i_m}^{im}(P_d) \\
&= \sum_{m=1}^{M} \frac{\partial U}{\partial R_{i_m}(P_d)} R_{i_m}(P_d)\lambda_{i_m}(P_d),
\end{aligned}
\tag{7.9}
$$

where $\{1, 2, \ldots, m, \ldots, M\}$ denotes the stepwise progression of node i from the fault state to the optimal operating state. $I_i^{im}(P_d)$ denotes the integrated importance of node i degrading to state 0 in different states. $I_{i_m}^{im}(P_d)$ denotes the integrated importance of node i in state m. U denotes the system performance function. $R_{i_m}(P_d)$ denotes the reliability of node i in state m. $\lambda_{i_m}(P_d)$ denotes the failure rate of node i degrading to state 0 in state m.

The direction of the maximum growth in system reliability can be derived from the representation of importance in the gradient. $\nabla \boldsymbol{f}$ denotes the gradient of the continuous function $f(x_1, x_2, \ldots, x_n)$. $\overrightarrow{\boldsymbol{x_1}}, \overrightarrow{\boldsymbol{x_2}}, \ldots, \overrightarrow{\boldsymbol{x_n}}$ denotes the unit orthogonal vector in the direction of x_i. The general expression for the gradient is as follows

$$
\nabla \boldsymbol{f} = \frac{\partial f}{\partial x_1}\overrightarrow{\boldsymbol{x_1}} + \frac{\partial f}{\partial x_2}\overrightarrow{\boldsymbol{x_2}} + \cdots + \frac{\partial f}{\partial x_n}\overrightarrow{\boldsymbol{x_n}}. \tag{7.10}
$$

The BIM of all nodes in a wind power system forms the gradient of the system reliability function $f(P_{w_1} = 1, P_{w_2} = 1, dots, P_{w_n} = 1)$, which indicates the direction at which the reliability of the wind power system increases fastest at the point $(P_{w_1} = 1, P_{w_2} = 1, \ldots, P_{w_n} = 1)$. The magnitude of the gradient determines the rate at which the reliability of the wind power system increases at that direction. The BIM of the node is expressed in the gradient as

$$
\begin{aligned}
\nabla \mathbf{Pr}\,\{\boldsymbol{\Phi}(\boldsymbol{P_w}) = \mathbf{1}\} &= I_1^B(P_d)\overrightarrow{\boldsymbol{Pr}_{\boldsymbol{P_{w_1}=1}}} + I_2^B(P_d)\overrightarrow{\boldsymbol{Pr}_{\boldsymbol{P_{w_2}=1}}} \\
&+ \cdots + I_n^B(P_d)\overrightarrow{\boldsymbol{Pr}_{\boldsymbol{P_{w_n}=1}}}.
\end{aligned}
\tag{7.11}
$$

The effect of node i on system reliability at different states of a wind power system forms the gradient of the system reliability function. The integrated importance of node i is the product of $\overrightarrow{\boldsymbol{v}}$ and the gradient of the system function is given by

$$
\begin{aligned}
\overrightarrow{\boldsymbol{v}} &= R_{i_1}(P_d)\cdot\lambda_{i_1}\cdot\overrightarrow{\boldsymbol{R_{i_1}(P_d)}} + R_{i_2}(P_d)\cdot\lambda_{i_2}\cdot\overrightarrow{\boldsymbol{R_{i_2}(P_d)}} + \ldots \\
&+ R_{i_M}(P_d)\cdot\lambda_{i_M}\cdot\overrightarrow{\boldsymbol{R_{i_M}(P_d)}}.
\end{aligned}
\tag{7.12}
$$

The geometric meaning of the integrated importance of node i in the gradient is the product of the projection of $\overrightarrow{\boldsymbol{v}}$ in the direction of the gradient of the reliability function of the wind power system and the magnitude of the gradient. The expression for the relationship between the IIM of node i and the gradient in a wind power system is

$$\begin{aligned}
&I_i^{im}(P_d) \\
&= \sum_{m=1}^{M} \frac{\partial U}{\partial R_{i_m}(P_d)} R_{i_m}(P_d)\lambda_{i_m}(P_d) \\
&= \left[R_{i_1}(P_d)\cdot\lambda_{i_1}, R_{i_2}(P_d)\cdot\lambda_{i_2}, \ldots, R_{i_M}(P_d)\cdot\lambda_{iM}\right] \\
&\cdot\left[\frac{\partial U}{\partial R_{i_1}(P_d)}, \frac{\partial U}{\partial R_{i_2}(P_d)}, \ldots, \frac{\partial U}{\partial R_{iM}(P_d)}\right]^{\mathrm{T}} \\
&= \vec{v}\cdot\left[\frac{\partial U}{\partial R_{i_1}(P_d)}\overrightarrow{\boldsymbol{R}_{i_1}(\boldsymbol{P}_d)} + \frac{\partial U}{\partial R_{i_2}(P_d)}\overrightarrow{\boldsymbol{R}_{i_2}(\boldsymbol{P}_d)} + \cdots + \frac{\partial U}{\partial R_{i_M}(P_d)}\overrightarrow{\boldsymbol{R}_{i_M}(\boldsymbol{P}_d)}\right] \\
&= \vec{v}\cdot\nabla R(P_d).
\end{aligned} \tag{7.13}$$

7.1.3 Case Study

The wind power systems shown in Fig. 7.1 is simplified into a logical network diagram for the wind power systems as shown in Fig. 7.3. The nodes in the logical network represent the generation nodes and the consumption nodes, and the edges represent the transmission network between the nodes. The logical network of the wind power systems is represented as $N(P, E)$, where P is the set of nodes and E is the set of edges. P contains two subsets: supply node subset PG and demand node subset PD. Supply node subset PG contains G1, G2, G5, G8, G11, and G13. Demand node subset PD contains D3, D4, D6, D7, and D9, D7, D9, D10, D12, D14, and D15 to D30.

Assuming that the nodes are independent of each other, the reliability of the demand node obeys the exponential distribution and the failure rate of the demand node is 0.006/week. The reliability of a supply node is assessed from the probability

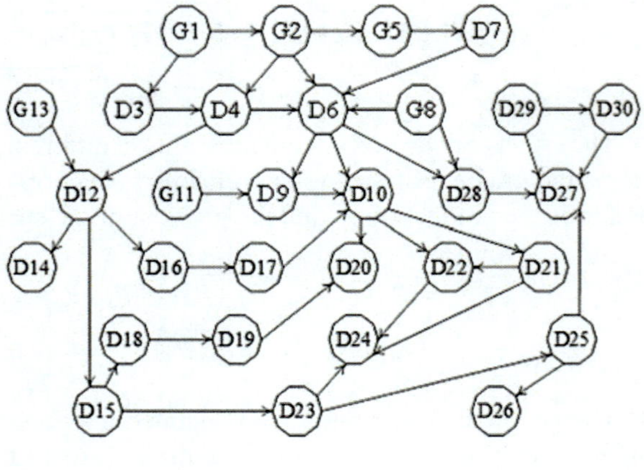

Fig. 7.3 Wind power logic network diagram

Table 7.1 The reliability of a supply node and a demand node

Supply node	Reliability	Demand node	Reliability
G1	0.92	D3, D4	0.6977
G2	0.92	D6, D7	0.6977
G5	0.92	D9, D10	0.6977
G8	0.92	D12	0.6977
G11	0.92	D14	0.6977
G13	0.92	D15–D30	0.6977

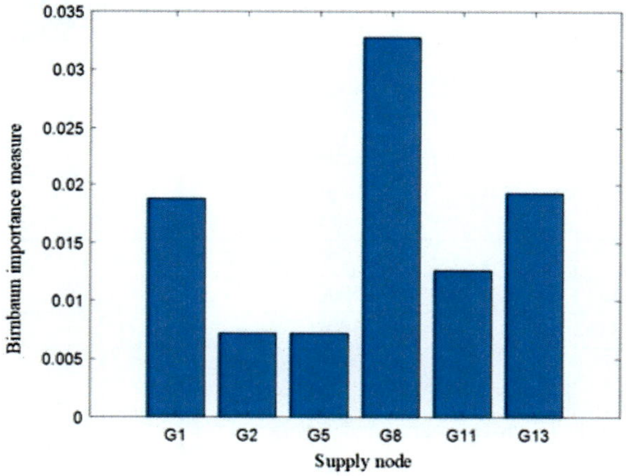

Fig. 7.4 The BIM of a supply node

of supply and demand electricity consumption, and the failure rate of the supply node is 0.004/week, 0.006/week, 0.005/week, 0.005/week, 0.008/week, and 0.007/week, respectively. Based on the monthly electricity consumption of 65 degrees in 2020, the demanded electricity consumption of a regional population per unit of time is 20 kw. The wind turbine operation cycle is 60 weeks, and the reliability of the node is $R_i(P_d)$. According to Eq. (7.3), the node reliability is shown in Table 7.1.

According to the wind power network diagram, the reliability of a supply node and a demand node, the reliability of the wind power systems are obtained. The degree of influence of the nodes on the reliability of the wind power system is evaluated by BIM and IIM. According to Eqs. (7.8) and (7.9), the impact of supply nodes on the wind power systems under BIM and IIM assessment is shown in Figs. 7.4 and 7.5.

According to BIM, the priority rankings of the impact of the supply node reliability on the reliability of wind power generation systems are G8, G13, G1, G11, G2, and G5, where G8 has the greatest impact on the reliability of wind power systems, and the state change of G8 causes the greatest change in the system reliability. According to the comprehensive importance, the priority rankings of the impact of a supply node reliability on the performance of wind power systems are G8, G13, G11, G1, G2,

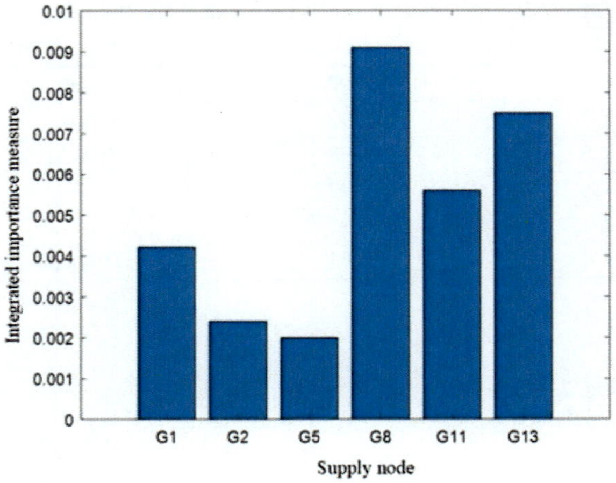

Fig. 7.5 The IIM of a supply node

Table 7.2 Supply node maintenance cost

Node	Maintenance cost/Yuan	Node	Maintenance cost/Yuan
G1	38000	G8	28000
G2	15000	G11	9000
G5	25000	G13	19000

Table 7.3 Maintenance node combinations

Maintenance node combinations	Nodes	Maintenance node combinations	Nodes
PG1	G1, G2, G11	PG6	G2, G5, G11, G13
PG2	G1, G5	PG7	G2, G8, G11
PG3	G1, G8	PG8	G2, G8, G13
PG4	G1, G11, G13	PG9	G5, G8, G11
PG5	G2, G5, G8	PG10	G8, G11, G13

and G5, where G8 has the greatest impact on the reliability of wind power systems, and the state change of G8 makes the greatest change of system reliability.

The wind turbine is the core of a wind power system. The severe operating environment of the wind power system can cause the wind turbine to fail and cause serious losses to the economy. Therefore, it is critical to perform repair on the failed wind turbine based on their reliabilities under limited resources. A typical distribution of a supply node repair costs is shown in Table 7.2.

Under the total maintenance cost constraint, the possible combinations of maintenance nodes are shown in Table 7.3.

The optimal orders of repair of the failed node combination PG1 with BIM and IIM evaluation are G1, G11, G2 and G11, G1, G2, respectively, and the reliability improvements of the wind power systems are 0.0386 and 0.0122, respectively. The optimal recovery order of the failed node combination PG2 under BIM and IIM evaluation are G1, G5, respectively; the reliability improvements of the wind power systems are 0.0260 and 0.0062, respectively. The optimal recovery orders of the failed node combination PG3 under the BIM and comprehensive importance evaluation are G8 and G1, respectively; the reliability improvements of the wind power systems are 0.0516 and 0.0133, respectively. The optimal recovery orders of the failed node combination PG4 under the BIM and comprehensive importance evaluation are 0.0516 and 0.0133, respectively. The optimal recovery orders of the failed node combination PG4 are G13, G1, G11 and G13, G11, G1 under BIM and IIM evaluation, respectively; and the reliability improvements of the wind power systems are 0.0507 and 0.0173, respectively. The optimal recovery order of the failed node combination PG5 is G8, G2, G5 under BIM and IIM evaluation, and the reliability improvement of the wind power systems are 0.0472 and 0.0135, respectively. The optimal recovery orders of the failed node combination PG6 under the BIM and comprehensive importance evaluation are G13, G11, G2, G5, and the reliability improvement of the wind power systems is 0.0463 and 0.0175, respectively. The optimal recovery orders of the failed node combination PG7 under the BIM and comprehensive importance evaluation are G8, G2, G5, and the reliability improvement of the wind power systems is 0.0463 and 0.0175, respectively. The optimal recovery orders of the failed node combination PG7 under both BIM and IIM evaluation are G8, G11, G2, and the reliability improvements of the wind power systems are 0.0526 and 0.0171, respectively; the optimal recovery orders of the failed node combination PG8 under both BIM and IIM evaluation are G8, G13, G2, and the reliability improvement of the wind power systems are 0.0593 and 0.0190, respectively; the reliability improvement of the failed node combination PG9 under BIM and comprehensive importance are G8, G11 and G5, and the reliability improvement of wind power systems are 0.0526 and 0.0167, respectively; the optimal recovery order of failed node combination PG10 under BIM and comprehensive importance are G8, G13 and G11, respectively and the reliability improvement of wind power systems is 0.0647. Therefore, the fault node maintenance order that maximises the reliability improvement of the wind power systems under the fixed cost constraint are G8, G13, and G11, respectively.

The supply node has a significant impact on the generation rate of the wind power systems and the demand node has a significant impact on the electricity consumption rate of the wind power systems. Supply nodes G1 and G8 and demand node G3 are selected as independent variables, respectively, and the reliability function of the wind power systems is $R = f(R_1, R_3, R_8)$. The reliability of G1 is $R_1 = 0.9200$, the reliability of G3 is $R_3 = 0.6977$, and the reliability of G8 is $R_8 = 0.9200$. The equivalence surface of the wind power system reliability is $R = f(R_1, R_3, R_8) = 0.9502$. According to Eq. (7.8), the BIM gradient of the wind power systems can be obtained by

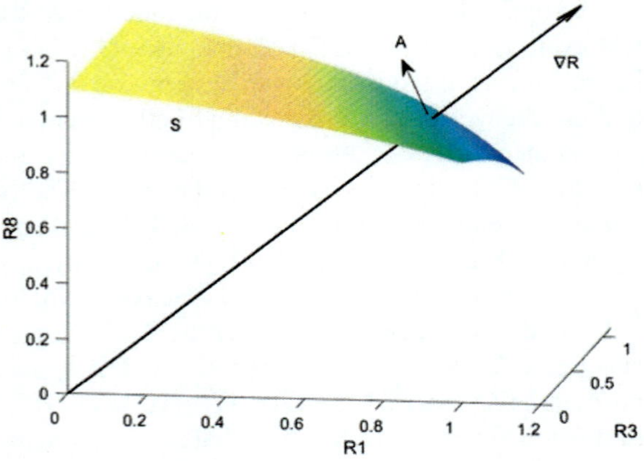

Fig. 7.6 BIM gradient of wind power systems

$$\nabla f(\mathbf{R_1}, \mathbf{R_3}, \mathbf{R_8}) = 0.0292\overrightarrow{\mathbf{R}}_1 + 0.0292\overrightarrow{\mathbf{R}}_3 + 0.0301\overrightarrow{\mathbf{R}}_8.$$

The BIM gradient of the wind power systems changes as shown in Fig. 7.5.

The surface S has the equivalent surface $R = f(R_1, R_3, R_8) = 0.9502$ for the reliability of the wind power systems. The vector $\nabla \boldsymbol{R}$ is the gradient $\nabla \boldsymbol{f}(\boldsymbol{R_1}, \boldsymbol{R_3}, \boldsymbol{R_8}) = 0.0292\overrightarrow{\boldsymbol{R}}_1 + 0.0292\overrightarrow{\boldsymbol{R}}_3 + 0.0301\overrightarrow{\boldsymbol{R}}_8$. The vector is the normal to the surface S. The intersection of the two nodes is A. The BIM of the wind power systems at the point A is $(I_1^B, I_3^B) = \overrightarrow{\boldsymbol{OA}}$. The direction of the vector $\nabla \boldsymbol{R}$ is the direction in which the reliability of the wind power systems grows fastest. Therefore, the facility manager should make the reliability of G1, G3 and G8 converge to point A to ensure that the system reliability grows along the fastest direction, see Fig. 7.6.

G1 and G3 are chosen as independent variables, and there are four states $\{0, 1, 2, 3\}$ for both the wind power systems and the supply and demand nodes. The probabilities of node G1 in the non-failure states are $R_{1_1} = 0.8276$, $R_{1_2} = 0.0650$, $R_{1_3} = 0.0273$, respectively. The probabilities of node G3 in different states are $R_{3_1} = 0.5827$, $R_{3_2} = 0.0743$, $R_{3_3} = 0.0407$, respectively. The failure rates of node G1 are $\lambda_{1_1} = 0.008$, $\lambda_{1_2} = 0.006$, $\lambda_{1_3} = 0.005$, respectively. According to Eq. (7.13), $\overrightarrow{\boldsymbol{v}} = 0.3972\overrightarrow{\boldsymbol{R}}_{1_1} + 0.0234\overrightarrow{\boldsymbol{R}}_{1_2} + 0.0082\overrightarrow{\boldsymbol{R}}_{1_3}$, the performance of the wind power systems function is $U = f(R_{1_1}, R_{1_2}, R_{1_3})$. The performance of the wind power systems equivalence surface is $U = f(R_{1_1}, R_{1_2}, R_{1_3}) = 71.3648$. According to Eq. (7.13), the IIM gradient of the wind power systems is obtained as $\nabla \boldsymbol{U} = 76.747\overrightarrow{\boldsymbol{R}}_{1_1} + 84.797\overrightarrow{\boldsymbol{R}}_{1_2} + 85.611\overrightarrow{\boldsymbol{R}}_{1_3}$. The variation of the IIM gradient of the wind power systems is shown in Fig. 7.7.

The vector $\nabla \boldsymbol{U}$ is the gradient $\nabla \boldsymbol{f}(\boldsymbol{R_{1,1}}, \boldsymbol{R_{1,2}}, \boldsymbol{R_{1,3}}) = 76.747\overrightarrow{\boldsymbol{R}}_{1_1} + 84.797\overrightarrow{\boldsymbol{R}}_{1_2} + 85.611\overrightarrow{\boldsymbol{R}}_{1_3}$. The vector $\nabla \boldsymbol{U}$ is normal to the surface Q. The intersection of the two is B. The projection point of vector $\overrightarrow{\boldsymbol{v}}$ on gradient $\nabla \boldsymbol{U}$ is C. The IIM of G1 at point B is $I_1^{im} = \overrightarrow{\boldsymbol{OB}} \cdot \overrightarrow{\boldsymbol{OC}}$. In a polymorphic system, the manager

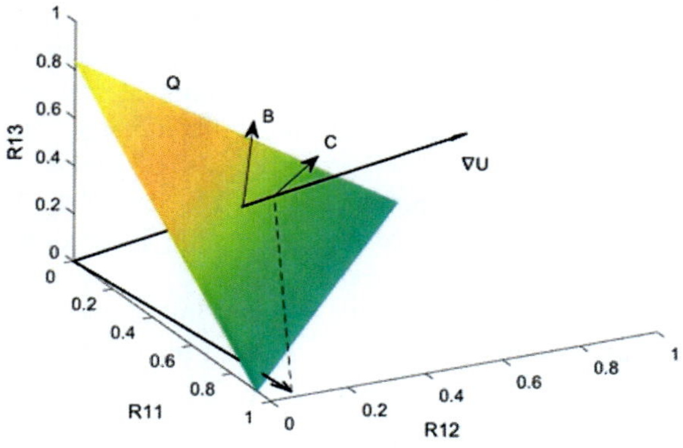

Fig. 7.7 IIM gradient of wind power systems

should make the probability of G1 in different states converge to point B to ensure that the system reliability grows along the fastest direction.

7.2 Satellite Attitude Control System

A satellite attitude control system composed of attitude sensors is a key subsystem of a satellite. Its feedback channel controls the satellite attitude by changing the control torque imposed on the attitude controller. If it fails, the consequences can be catastrophic. By estimating the probability of failures, engineers can propose targeted preventive maintenance (PM) policies to improve the reliability of the system. This section focuses on the PM of the satellite attitude control system.

PM is an effective method to extend the reliability of a system. Geng et al. [17] proposed a satellite fault diagnosis and evaluation method based on virtual maintenance, with the aim to reduce the subjectivity of satellite assembly fault diagnosis and evaluation for factors such as accessibility, ergonomics, wiring and the damage degree in satellite assembly fault diagnosis. Assembly or troubleshooting schemes can be improved at an early stage of satellite design. Zhao et al. [28] took mission-critical systems as the research objective and optimised maintenance policies and performance control strategies jointly based on workload and system states, so as to balance the mission success probability and system survivability. Perrons and Richards [22] suggest that the upstream oil and gas industry can make a significant progress in asset management by selectively applying some aspects of maintenance policies and concepts learned in the space and satellite sectors, particularly offshore platforms and long-range pipelines. Zhu et al. [30] proposed an importance measure by considering various maintenance effectiveness such as imperfect maintenance and replacement, and proposed a PM policy by considering the residual life and residual profit. For systems that fail due to degradation or external factors, Hashemi et al. [18] proposed an optimal PM model considering the service age and PM cost, corrective maintenance (CM) cost and minimum maintenance cost. Through an in-depth discussion of the impact of maintenance on a structural performance function, Shi et al. [23] established an optimisation model of a PM policy for a system. Finkelstein et al. [15] were the first to discuss the optimal maintenance of dynamic groups of statistically identical items in the reliability literature. Wei et al. [25] proposed a method to describe degrade stochastic processes and analysed the complex trade-off between state-based PM and buffer capacity. Nasrfard et al. [21] proposed a Petri net maintenance model that considered degradation, inspection and maintenance processes, as well as random and age-related faults.

To avoid serious economic losses, this section develops a PM policy for satellite attitude control systems by using the importance measures proposed in Chap. 3.

7.2.1 Degradation Modelling in External Shocks

The criticality of components usually depends on the degree of degradation, the level of fault threshold caused by degradation, dynamic environmental conditions, and satellite attitude control system configuration.

Assume that a satellite attitude control system is composed of n components. The degradation process j of component i is denoted as $\left(X_j^{(i)}\right)_{j=1,2,\ldots,k}$. When any degradation process k_i relating to this component reaches the threshold level $n_j^{(i)}$, component i fails. That is, component i has multiple competing failure modes due to a multi-dimensional degradation. Taking the momentum wheel component as an example, there are five failure modes, which are: stuck fault, idling fault, friction fault, gain reduction fault, and jump fault. When any of the five degradation processes reaches a certain threshold level, the component fails.

The degradation process of the satellite attitude control system components is modelled by a multi-dimensional Wiener process. Based on the definition of environmental importance given in Chap. 3, the relationship between the degradation process and external shocks can be established, and the PM policy based on environmental importance can be further obtained. The details will be further given in the case study in Sect. 7.2.2.

7.2.2 Case Study

The failure mode of different components of a satellite attitude control system is different. The main components of a satellite attitude control system are the momentum wheel, the star sensor, the sun sensor and the infrared earth sensor, among which the momentum wheel has five failure modes, namely: stuck fault, idling fault, friction fault, gain reduction fault, and jump fault. The star sensor, the sun sensor and the infrared earth sensor have three failure modes: stuck fault, deviation fault, and noise fault. Table 7.4 shows the major failure modes of each component.

Table 7.4 The main failure modes of satellite attitude control system components

No.	Failure mode name	No.	Failure mode name
A1	Stuck fault	B3	Noise fault
A2	Idle fault	C1	Stuck fault
A3	Friction fault	C2	Deviation fault
A4	Gain drop fault	C3	Noise fault
A5	Jump fault	D1	Stuck fault
B1	Stuck fault	D2	Deviation fault
B2	Deviation fault	D3	Noise fault

Table 7.5 Parameter values of simulation

Failure mode	$\mu_{j,0}^{(i)}$	$\sigma_{j,0}^{(i)}$	Initial degradation level $x_j^{(i)}(0)$	$n_j^{(i)}$
Momentum wheel				
A1	1	0.5	0	11
A2	1	0.8	0	11
A3	0.8	1	0	11
A4	0.8	0.7	0	11
A5	1	1	0	11
Star sensor				
B1	0.8	0.5	1	12
B2	0.7	1	1	12
B3	0.6	1	1	12
Sun sensor				
C1	0.7	1	4	15
C2	0.9	0.8	4	15
C3	1	0.5	4	15
Infrared earth sensor				
D1	1.1	0.7	2	13
D2	1.3	0.9	2	13
D3	0.6	1.3	2	13

Assuming that the environmental condition is determined and fully known before t =30, the external shocks e_t is expressed by the piecewise function as:

$$e_t = \begin{cases} 6, \ 0 \le t < 10 \\ 7, \ 10 \le t < 30 \end{cases}. \tag{7.14}$$

The specific parameter values of the satellite attitude control system used in this example are shown in Table 7.5.

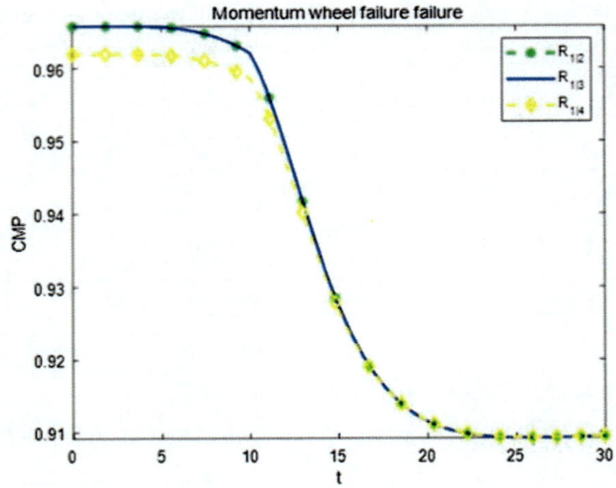

Fig. 7.8 Momentum wheel failure

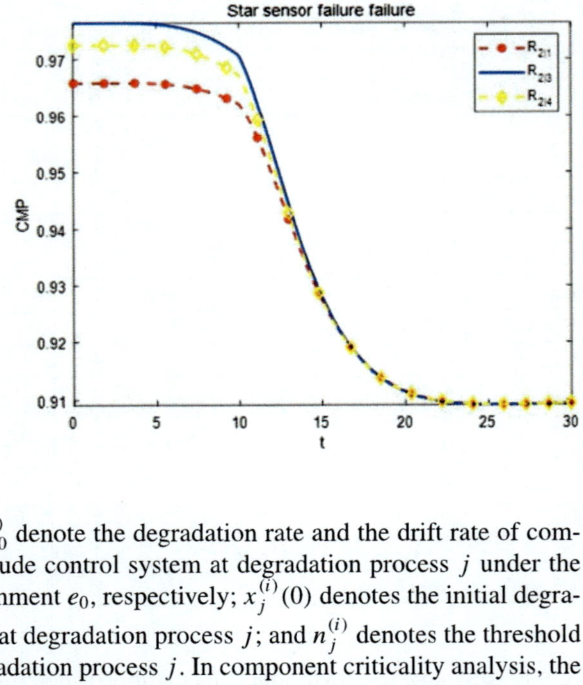

Fig. 7.9 Star sensor failure

In Table 7.5, $\mu_{j,0}^{(i)}$ and $\sigma_{j,0}^{(i)}$ denote the degradation rate and the drift rate of component i in the satellite attitude control system at degradation process j under the condition of constant environment e_0, respectively; $x_j^{(i)}(0)$ denotes the initial degradation level of component i at degradation process j; and $n_j^{(i)}$ denotes the threshold level of component i at degradation process j. In component criticality analysis, the system is assumed to be a repairable system and its components are not new. In this example, it is assumed that the initial degradation levels of the star sensor, the sun sensor, and the infrared earth sensor are greater than zero.

The PM policy model is simulated and analysed, and the parameters of each component are analysed to obtain the PM priority on the other components in case of a component failure. The simulation results are shown in Figs. 7.8, 7.9, 7.10 and 7.11.

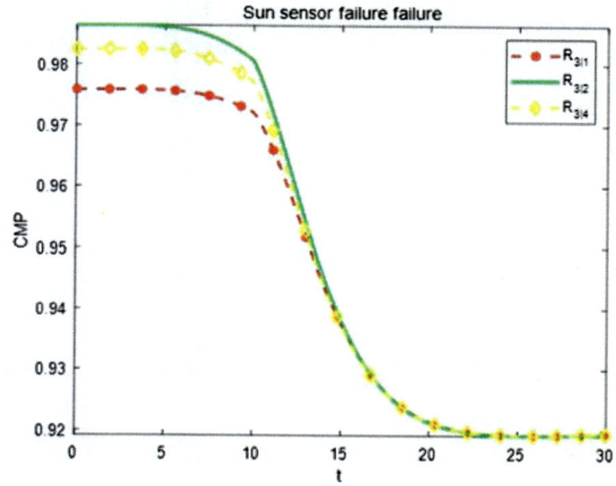

Fig. 7.10 Sun sensor failure

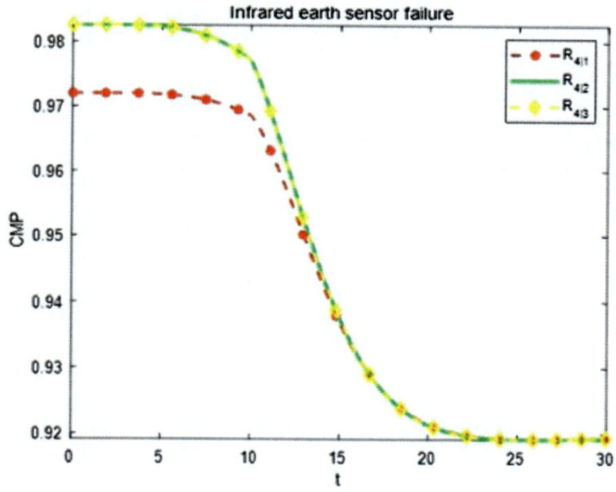

Fig. 7.11 Infrared earth sensor failure

As shown in Figs. 7.8, 7.9, 7.10 and 7.11, when a certain component degrades below the threshold value, the priority of PM of other components will first remain unchanged, then decline and become flat. In addition, the CMP values of the servo-motor and the servo-driver are almost the same as those during the period $0 \leq t < 30$, and the CMP values of the star sensor and the sun sensor are almost the same, and the order of CMP is higher than that of other components. The results of the analysis provide a comprehensive ranking of the priority for PM of the remaining components in the event of a component failure, as shown Table 7.6.

After determining the overall CMP ranking of each component, some cost control due to maintenance and PM is taken into account. Different total maintenance costs may lead to different PM programs. Table 7.7 shows the maintenance and PM costs of major components of the satellite attitude control system.

Table 7.6 Cropland node number table

	Momentum wheel	Star sensor	Sun sensor	Infrared earth sensor
Momentum wheel	–	1	2	3
Star sensor	3	–	1	2
Sun sensor	3	1	–	2
Infrared earth sensor	4	1	2	–

Table 7.7 PM costs of major components of the satellite attitude control system

No.	Component	Corrective maintenance cost	CM cost
1	Momentum wheel	\$7000	\$2200
2	Star sensor	\$4000	\$1800
3	Sun sensor	\$3900	\$1700
6	Infrared earth sensor	\$2600	\$1400

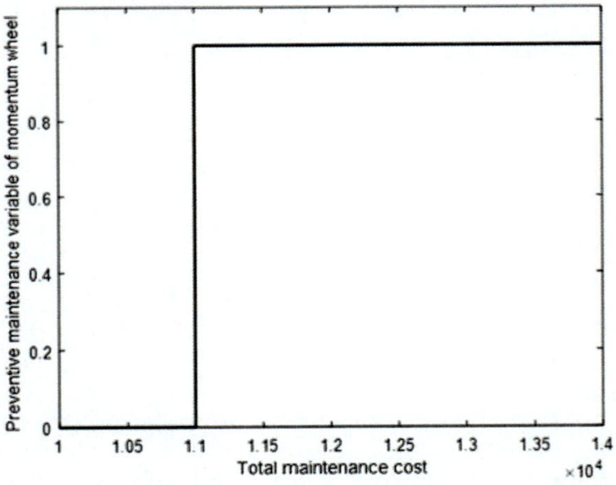

Fig. 7.12 Total maintenance cost of momentum wheel

As can be seen from Table 7.7, when a component fails at different times t, the set of components for PM are selected under different total maintenance cost constraints. Taking the momentum wheel failure at $t = 10$ and PM of other components as an example, the simulation results are shown in Figs. 7.12, 7.13, 7.14 and 7.15.

As can be seen from Figs. 7.12, 7.13, 7.14 and 7.15, combined with the constraints of PM cost and component maintenance priority, it can be seen that although the PM cost of the star sensor and the sun sensor is higher, the priority of components for PM is higher, so no matter how the total maintenance cost changes, these two

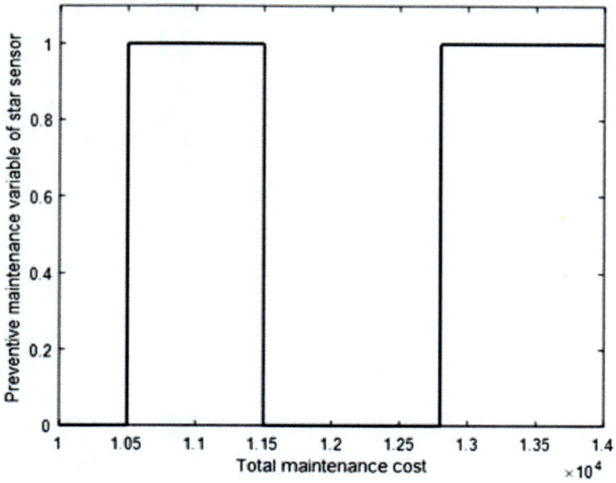

Fig. 7.13 Total maintenance cost of star sensor

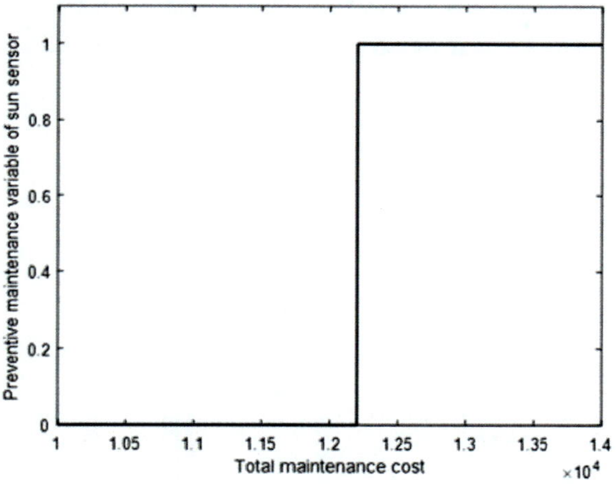

Fig. 7.14 Total maintenance cost of sun sensor

components are preferred for PM. In addition, although the component repair priority of the infrared Earth sensor is lower, the lower PM cost means that if the remaining maintenance cost is not enough to repair other priority components, the infrared Earth sensor is selected for PM.

Taking the failure of the retarder at $t = 10$ and the total maintenance cost of 11200 Renminbi Yuan as an example, the influence of this PM policy on system reliability was analysed after selecting the set of components: the star sensor and the sun sensor, for PM. Figures 7.16 and 7.17 show the reliability curve of each component with running time before and after PM.

When $t = 10$, the reliabilities of each component are shown in Table 7.8. The comparison shows that the PM policy plays a significant role in improving system reliability.

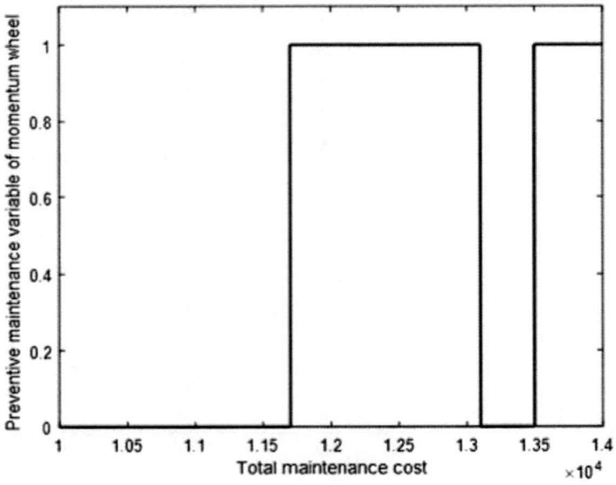

Fig. 7.15 Total maintenance cost of infrared earth sensor

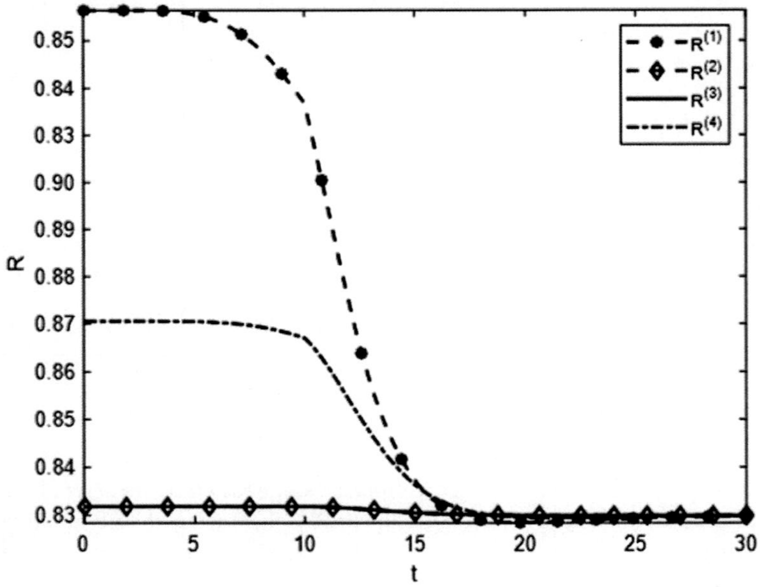

Fig. 7.16 Reliability changes of components before PM

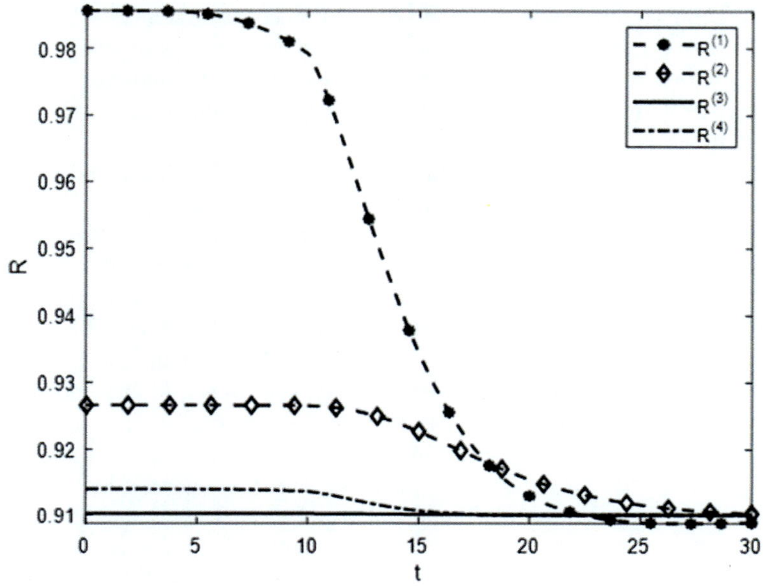

Fig. 7.17 Reliability changes of components after PM

Table 7.8 Reliability value of each component when $t = 10$

Component reliability	Before PM	After PM
Momentum wheel R(1)	0.847	0.978
Star sensor R(2)	0.831	0.927
Sun sensor R(3)	0.830	0.910
Infrared earth sensor R(4)	0.838	0.914

7.3 Rocket Vertical Assembly and Test Plant System

As one of the subsystems of China's space engineering, the system of the space launch site is an important component of the space engineering, and it is the starting point for successfully sending spacecrafts into the space. The system of the space launch site mainly consists of the launch zone and the technical zone. The launch zone is mainly responsible for the launch of rockets and a spacecraft, and its main ground facilities are the launch towers, while the technical zone is mainly responsible for the testing and assembly of rockets and spacecrafts, and the abbreviation rocket vertical assembly and test plant (RVATP) is the main ground facilities of the technical zone. The RVATP is an important place for general assembly and testing of rockets. The platform hydraulic system, the vertical transfer gate, the sliding gate, the double trolley bridge crane, the lifting work platform, the movable table auxiliary platform, the cable auxiliary platform, and other key equipment or systems such as the fire protection system and the air conditioning system are the keys to ensure a normal operation of this system. Due to the complex composition of the above equipment and the special working environment, the mechanical, hydraulic and electrical systems are prone to fail.

If a non-critical component fails, the system does not shut down, and PM can be performed on other non-critical components at this time [2]. Appropriate PM can improve the reliability and quality of other components and reduce the probability of future failures of that component, thereby improving the overall reliability of the system [4]. Gao et al. [16] considered two maintenance policies, namely PM policies at the end of each production cycle and scheduling PM policies at each set point. Dui et al. [11] proposed a cost-based prioritisation approach for PM on multi-state systems, based on the buffer capacity, and discussed three maintenance policies. Importance measures in reliability engineering have the advantage of easily identifying weak points in a system and characterising the impact of components on the system, providing valuable information for system maintenance, and thus is widely used in the selection of components for PM [10]. Dui et al. [9] extended some important measures to answer the question of component selection for PM when CM is also applied. In addition, considering the cost and time of component repair, Dui et al. [8] proposed a comprehensive cost-based importance measure to identify components or groups of components that can be selected for PM.

Based on the cost-based component maintenance importance introduced in Chap. 4, a maintenance cost index is proposed to measure the maintenance priority of components. The index is applied to the RVATP system.

7.3.1 Fault Analysis of Rocket Vertical Assembly and Test Plant System

The most undesired event of the whole system is the failure of the RVATP, which is the top event, while its intermediate events are the failures of some subsystems such as the platform hydraulic system, the vertical transfer gate, the sliding gate, the double trolley bridge crane, the air conditioning system, and the fire protection system. Each intermediate event in those subsystems can continue decomposing down as a top event to build the fault tree of each subsystem separately.

Figures 7.18, 7.19, 7.20, 7.21, 7.22, 7.23 and 7.24 show the fault trees associated with the RVATP. By analysing these fault trees, Table 7.9 can be obtained. Table 7.9 shows that each subsystem in the RVATP is the top-level event of the sub-fault tree, and these subsystems can continue to be decomposed down to the component level. Intermediate events B01 to B18 represent different component failure events, respectively. Failure of any one of these components can lead to degraded performance or even failure of the RVATP. In order to ensure that the RVATP is up to the task, the important components corresponding to B01 to B18 need to be subjected to PM.

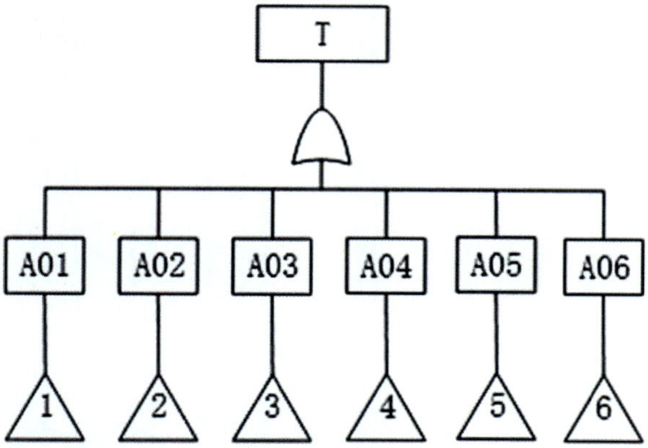

Fig. 7.18 Reliability changes of components after PM

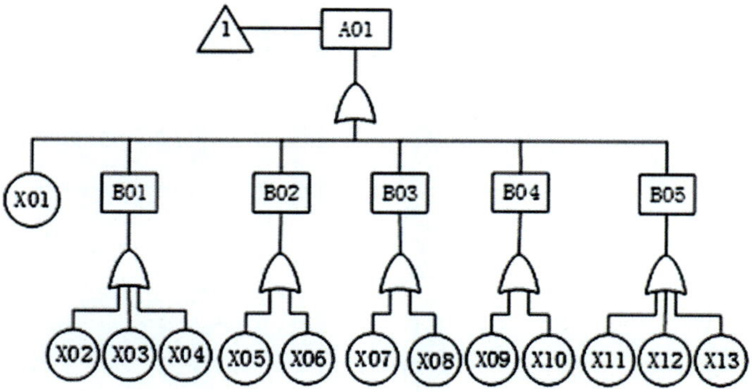

Fig. 7.19 Fault tree of the RVATP

Fig. 7.20 Fault tree of a platform hydraulic system

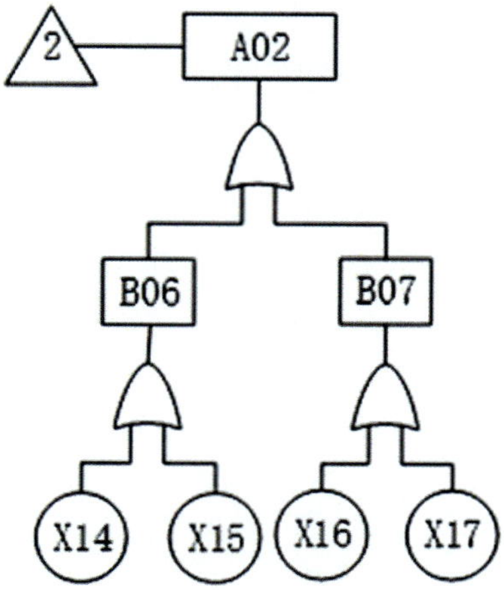

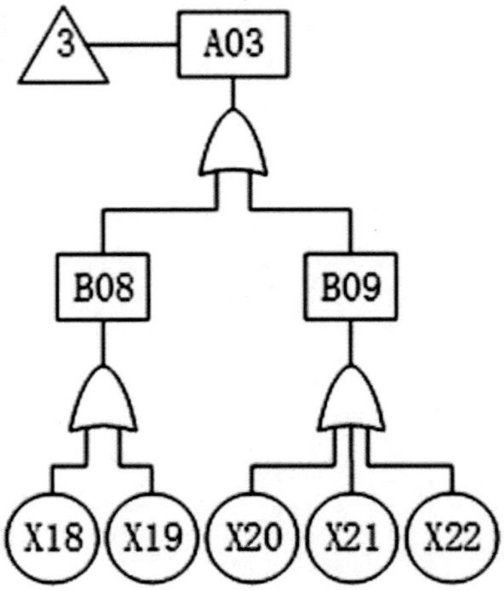

Fig. 7.21 Fault tree of a vertical transfer gate

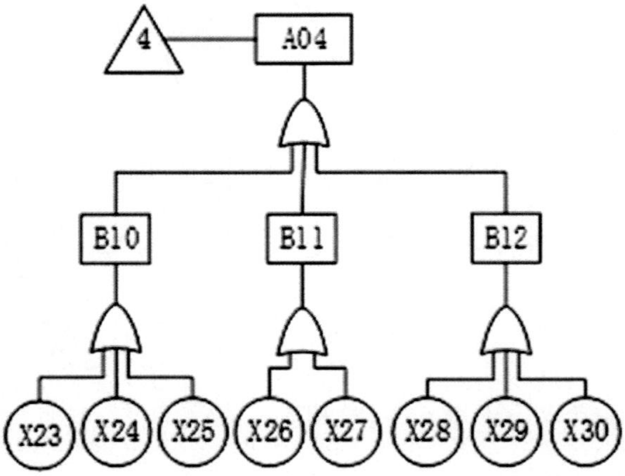

Fig. 7.22 Fault tree of a sliding door

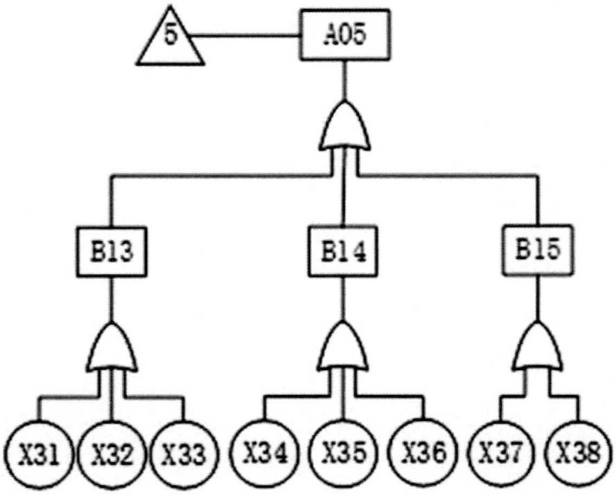

Fig. 7.23 Fault tree of a double trolley bridge crane

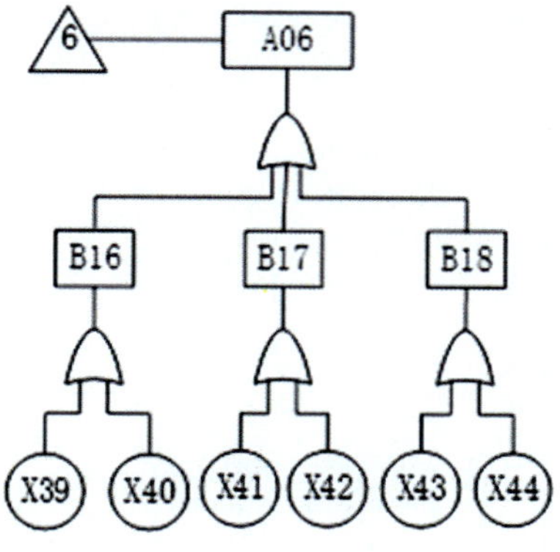

Fig. 7.24 Fault tree of an air conditioning system

The Weibull distribution is widely used in reliability engineering, and is used in modelling wear accumulation failures. We assume that the above 18 important components obey the Weibull distribution $W(t;\, \theta, \gamma)$, where θ is the scale parameter and γ is the shape parameter. The reliability function of the component q can be obtained as $R_q(t) = \exp[-(\frac{t}{\theta})^{\gamma-1}]$ and the failure rate as $\lambda_q(t) = \frac{\gamma}{\theta}(\frac{t}{\theta})^{\gamma-1}$. The reliability function and failure rate function of the components will be applied in the case studies.

The values of θ and γ are shown in Table 7.10, respectively.

The repair costs and PM costs for each important component can be estimated from some historical data. These costs are given by Table 7.11.

The various types of failure modes derived in the fault tree analysis have different degrees of impact on the overall system performance. As a result, the state of the system is classified according to the frequencies of the occurrences of the failure modes. The performance indicators of the components corresponding to these states are given. There are 46 states. States 1–44 are the intermediate states of the RVATP, state 45 is the complete failure state, and state 46 is the perfect operation state, as detailed in Table 7.12.

7.3.1.1 Total Expected Cost

In selecting components for PM, the goal must be to minimise the expected total maintenance cost value of the RVATP. In this section, the expected total maintenance cost of the system is derived for the specific scenarios, where three different maintenance costs are considered.

Once a component failure is detected, the maintenance team will repair it immediately. Only the failure of critical components will cause the failure of the RVATP. If the failed component is a critical component and incurs system downtime costs, PM can be performed on other components. If the failed component is not a critical component, the failure of this component will not cause the RVATP to shut down, and PM activities can then be performed on other non-critical components that do

Table 7.9 Fault tree event codes and names of the RVATP

Code	Name	Code	Name
T	Rocket vertical assembly test plant failure	X11	High load pressure of each cylinder
A01	Platform hydraulic system failure	X12	Abnormal vibration of each cylinder
A02	Vertical transit gate failure	X13	Abnormal hydraulic oil flow of each cylinder
A03	Sliding door failure	X14	Solenoid valve coil failure
A04	Double trolley bridge crane failure	X15	Solenoid valve spool failure
A05	Air conditioning system failure	X16	Throttle flow adjustment failure
A06	Fire protection system failure	X17	Increased leakage in the throttle valve
B01	Hydraulic pump failure	X18	Abnormal position of the main door
B02	Electric motor failure	X19	Abnormal vibration of the gate body
B03	Proportional valve failure	X20	Abnormal variable frequency motor power
B04	Hydraulic motor failure	X21	Abnormal vibration of inverter motor
B05	Each platform cylinder failure	X22	Inverter motor case with electricity
B06	Solenoid valve failure	X23	Abnormal variable frequency motor power
B07	Throttle valve failure	X24	Abnormal vibration of inverter motor
B08	Solenoid valve failure	X25	Inverter motor case with electricity
B09	Throttle valve failure	X26	Lack of brake fluid in the brake
B10	Inverter motor failure	X27	Brake pad wear
B11	Brake failure	X28	Abnormal speed reducer vibration
B12	Reducer failure	X29	Reducer oil seal leakage
B13	Compressor failure	X30	Abnormal output pressure of reducer oil pump
B14	Blower failure	X31	Compressor blocking and abnormal load
B15	Ventilation duct failure	X32	Abnormal compressor vibration and shaft displacement
B16	Fire hydrant failure	X33	Insufficient cooling
B17	Flue gas control failure	X34	Blower motor overload
B18	Fire protection network failure	X35	Insufficient blower flow
X01	Abnormal hydraulic oil temperature	X36	Blower abnormal vibration and noise
X02	Abnormal hydraulic pump oil discharge volume	X37	Poorly connected air ducts
X03	Hydraulic pump output pressure failure	X38	Abnormal vibration and noise
X04	Abnormal noise or vibration of hydraulic pump	X39	Hydrant pressure stabilization failure
X05	Abnormal motor vibration	X40	Fire pump power failure
X06	Motor power failure	X41	Smoke exhaust valve starts abnormally
X07	Abnormal displacement of proportional valve spool	X42	Abnormal power of range hoods
X08	Proportional valve filter port clogged	X43	Abnormal vibration of fire protection pipe network
X09	Excessive load pressure of hydraulic motor	X44	Leakage caused by failure of pipe network seals
X10	Hydraulic motor outlet flow abnormal		

Table 7.10 Parameters related to component failure time and repair time

Code	θ_{1i}	γ_{1i}	θ_{2i}	γ_{2i}
B01	2045	2.43	8	2
B02	4385	1.95	10	2
B03	3015	2.24	4	3
B04	2045	2.43	4	2
B05	3364	1.21	6	2
B06	3015	2.24	2	3
B07	3015	2.24	7	2
B08	3015	2.24	12	3
B09	3015	2.24	15	2
B10	4385	1.95	4	2
B11	3207	2.11	8	2
B12	3207	2.11	8	2
B13	3321	1.97	7	2
B14	3532	2.01	14	3
B15	1722	2.12	12	2
B16	2249	1.44	15	2
B17	3159	2.17	12	2
B18	1722	2.12	15	3

Table 7.11 Costs associated with RVATP

Code	ε_q	$\varepsilon_{q,PM}$	Code	ε_q	$\varepsilon_{q,PM}$
B01	29875	15531	B10	33157	21364
B02	33157	21364	B11	16814	13593
B03	15211	10468	B12	16814	13593
B04	29875	15531	B13	21357	17436
B05	11855	8264	B14	19412	13827
B06	15211	10468	B15	12864	8673
B07	15211	10468	B16	12733	9431
B08	15211	10468	B17	20415	16933
B09	15211	10468	B18	12864	8673

not constitute a cut set and only incur cost of repairing the component and cost of PM on the other components. Denote $c(t)$ as the expected total maintenance cost of the RVATP, ε_s^q as the cost of system downtime due to the failure of component q, and $\varepsilon_{PM}^q(t)$ as the cost of PM of other components after the failure of component q. Since the failed component q may be a critical or non-critical component, two expressions are subsequently used to discuss both cases. According to Eq. (4.3), in addition to the PM cost, the expected total maintenance cost of the system over the time interval $(0, t)$ can be given by

Table 7.12 System status and corresponding performance index

No.	State	Performance index
1	X01	Hydraulic oil temperature
2~4	X02/X03/X04	Hydraulic pump output pressure
5~6	X05/X06	Motor output power
7~8	X07/X08	Proportional valve output flow
9~10	X09/X10	Hydraulic motor output pressure
11~13	X11/X12/X13	Vibration of the cylinder
14~15	X14/X15	Solenoid valve control precision
16~17	X16/X17	Throttle valve control precision
18~19	X18/X19	Gate displacement
20~22	X20/X21/X22	Motor output stability
23~25	X23/X24/X25	Motor output stability
26~27	X26/X27	Restraining driving force
28~30	X28/X29/X30	Matching precision
31~33	X31/X32/X33	Lifting pressure
34~36	X34/X35/X36	Conveying gas medium
37~38	X37/X38	Drafting efficiency
39~40	X39/X40	Fire hydrant output
41~42	X41/X42	Flue gas removal efficiency
43~44	X37/X38	Leakage
45	0	Reliability
46	1	Reliability

$$c(t) = \sum_{q=1}^{m} \{ \left\{ \varepsilon_s^q p_{s(<K)}^q(t) + \varepsilon_q + \varepsilon_{PM}^q(t) \right\} p_0^q(t) \}, \tag{7.15}$$

where m is the number of components in the RVATP; $p_{s(<K)}^q(t)$ is the probability that the system performance is below the threshold K, caused by the failure of component q at time t. ε_q is the repair cost of the failure of component q, and $p_0^q(t)$ is the probability of the failure of component q at time t. The state x_q of component q can be expressed as $x_q(t) = 1$ if component q is working at time t and $x_q(t) = 0$ otherwise. Denote by $\phi(\boldsymbol{X}(t))$ the structure function of the RVATP at the time t whose value domain is $\{0, 1\}$ and $\phi(\boldsymbol{X}(t)) = \phi(x_1(t), x_2(t), \cdots, x_m(t))$. For example, $(0_q, 1_{\neq q})$ means that component q fails and the other components work normally. $p_{s(<K)}^q(t)$ can be expressed as $\Pr\left[\phi(0_q, 1_{\neq q}) < K\right]$, i.e., the probability that the system performance is less than the threshold K when component q fails and the other components are working normally. $p_0^q(t)$ can be expressed as $\Pr\left[x_q(t) = 0\right]$, i.e., the probability of failure of component q. The reliability of component q can also be expressed as $p_q(t)$ and $p_q(t) = \Pr\left[T_q > t\right]$, with T_q being the life time of component q.

Assume that component q is a critical component. If component q fails and then the RVATP fails, the PM cost of other components at time t is

$$\begin{aligned}\varepsilon_{PM}^{q}(t) &= \sum_{l=1,l\neq q}^{m} \{\varepsilon_{l,PM} p_1^l(t) p_{s=0}^q(t)\} \\ &= \sum_{l=1,l\neq q}^{m} \{\varepsilon_{l,PM} \Pr[x_l(t)=1]\},\end{aligned} \tag{7.16}$$

where $\varepsilon_{l,PM}$ is the PM cost of component l. When PM is performed on component l, component l must stop working. $\Pr\left[\phi(0_q, 0_l, 1_{\neq q,l}) = 0\right]$ represents the probability that the RVATP is in a shutdown state when components q and l stop working and the other components do not stop working. Since component q is a critical component, the system is shut down. At this point, it is not considered that PM on other components will affect the operation of the RVATP.

If the failed component q is a non-critical component, the RVATP does not fail. Some components in the system are then selected for PM because PM cannot be performed on the critical components. If PM is performed for a non-critical component l, there should be two requirements for this component: i.e., component l is not a critical component, and component l and the failed component q do not form a cut set.

Assume that component q is a non-critical component. If it fails, the PM cost of the other components at time t is

$$\varepsilon_{PM}^{q}(t) = \sum_{l=1,l\neq q}^{m} \{\varepsilon_{l,PM} \Pr[x_l(t)=1] \Pr\left[\phi(0_q, 0_l, 1_{\neq q,l}) \geq K\right]\}, \tag{7.17}$$

where $\Pr\left[\phi(0_q, 0_l, 1_{\neq q,l}) \geq K\right]$ is the probability that the system remains under the normal operation when component q and component l fail and the other components are working normally. The RVATP operates normally and does not incur system downtime cost.

7.3.1.2 Maintenance Policies

In the practical application of the RVATP, regardless of whether the component is critical or non-critical, the repair team need to maintain the component and trigger PM immediately upon its failures. Other components can perform PM at the same time.

It is assumed that a component can be immediately identified once it fails. When a component's failure is identified, repair is performed immediately and PM can be performed on other non-faulty components simultaneously. When a critical component fails, PM can be performed on all other components. When a non-critical component fails, since this strategy can perform PM measures on other components simultaneously, it is necessary to identify whether other non-critical components form a cut set with that component to prevent failures in the RVATP. The total expected cost function for the RVATP in case of failure of non-critical components needs to be given by

$$c_S(t) = \sum_{q=1}^{m} \{ \left\{ \varepsilon_s^q p_{s(<K)}^q (t) + \varepsilon_q + \varepsilon_{PM}^{S,p} (t) \right\} p_0^q (t) \}, \tag{7.18}$$

$$\varepsilon_{PM}^{S,p} (t) = \sum_{i=1}^{n} \{ \varepsilon_{l_i,PM} p_1^{l_i} (t) \, p_{s(\geq K)}^q (t) \}, \tag{7.19}$$

where $\varepsilon_{PM}^{S,p} (t)$ represents the PM cost to other components in the system after the failure of component q. The number of PMs that can be executed simultaneously in this system is n. $\varepsilon_{l_i,PM}$ represents the PM cost of component l_i. $p_1^{l_i} (t)$ represents the probability that component l_i works properly, which can be expressed in terms of the structure function as $\Pr \left[x_{l_i} (t) = 1 \right]$. $p_{s(\geq K)}^q (t)$ represents the probability that the system state is greater than the threshold K after component q's failure. The structure function can then be expressed as $\Pr \left[\theta \left(0_q, 0_{l_1}, 0_{l_2}, \cdots, 0_{l_i}, 1_{\neq q, l_1, l_2, \cdots, l_i} \right) \geq K \right]$. According to Eq. (4.3), we can obtain the CCMI of component l. In a practical application of a RVATP, it may be more common to consider specific cost constraints for the PM strategy. Therefore, to perform PM for other components one by one when one component fails, we need to solve the following integer planning problem. According to Eq. (4.3) and the influence of the system structure, the following integer plan can be obtained:

$$\max \theta \left(0_q, x_{l_1}, x_{l_2}, \ldots, x_{l_i}, 1_{\neq q, l_1, l_2, \cdots, l_i} \right) \sum_{l=1, l \neq q}^{m} I_l^c d_l, \tag{7.20}$$

subject to

$$\sum_{l=1, l \neq q}^{m} \varepsilon_{l,PM} \leq C.$$

where $\varepsilon_{l,PM}$ represents the PM cost of component l, C is the maintenance cost budget for the RVATP, d_l is a decision variable that can only have a value of 0 or 1 and is used to indicate whether to select the component for PM. The structure function $\theta \left(0_q, x_{l_1}, x_{l_2}, \cdots, x_{l_i}, 1_{\neq q, l_1, l_2, \cdots, l_i} \right)$ is used to determine that the selected PM component will not cause the RVATP to shut down.

7.3.2 *Case Study*

The composition of the RVATP is relatively complicating. From Figs. 7.18, 7.19, 7.20, 7.21, 7.22, 7.23 and 7.24, we can understand that the system has six subsystems, which mainly include the platform hydraulic system, the vertical transfer gate, the sliding gate, the double trolley bridge crane, the air conditioning system, and the fire protection system. Among all the mechanical systems, pumps, motors and electronic control systems are the devices with relatively high failure frequencies. Therefore, in the RVATP, the critical components are B01, B02, B04, B10, B14, and B17, and the other components are non-critical components. According to the composition structure of the RVATP, we can know that components B01 and B04 are of the same type of components, so their CCMI are the same. Similarly, B02 and B10 are motors of the same type, B03, B06, B07, B08, and B09 are all valve type components, B11 and B12 are both brakes, and B15 and B18 are networks of the same type. In the analysis hereafter, only one of these components is simulated due to the same maintenance cost, PM cost, and maintenance-related parameters for the same class of mechanical components.

The CCMI of a component relates to its maintenance costs. As maintenance costs increase, the CCMI increase. This means that it is more valuable to perform PM on a component when it fails and higher maintenance costs are then incurred. When a component fails, CCMI is related to the location of the component in the system. The impact of the component location of the PM on the CCMI value depends only on whether the component is a critical component. Below, a more specific analysis of CCMI for critical and non-critical component failure scenarios is shown in Figs. 7.25 and 7.26, respectively.

If a critical component fails, the RVATP fails so that PM can be performed on all other components. Figure 7.8 shows the variation of CCMI over time when a critical component fails. The CCMI is affected not only by the cost associated with its PM component but also by the reliability of the other components selected for PM. It is clear that the CCMIs of different components for PM selected are interleaved. It can be found that when the same critical component fails, the CCMI of all components for PM available for selection differ in size at the same time. The priority of its PM can be determined based on the size of the CCMI value. It is worth noting that the priority of components for PM changes over time. Furthermore, when the failed critical components are different, the CCMI and repair priorities of the same components for PM are also different. For example, when component B01 or B04 fails, the priority for PM on the other components at $t = 50$ are B02, B14, B03, B17, B15, B16, B11, B13, and B05. At $t = 90$, the priority for PM on the other components is ranked as B02, B03, B14, B15, B16, B17, B11, B13, and then B05. The results of other critical components are shown in Fig. 7.25.

Figure 7.26 shows the change in the CCMI corresponding to the components selected for PM when the non-critical component fails. When non-critical components B03, B05, B11, B13, B15, and B16 fail, PM cannot be performed on critical components since the system is not failed. At this time, PM can only be performed

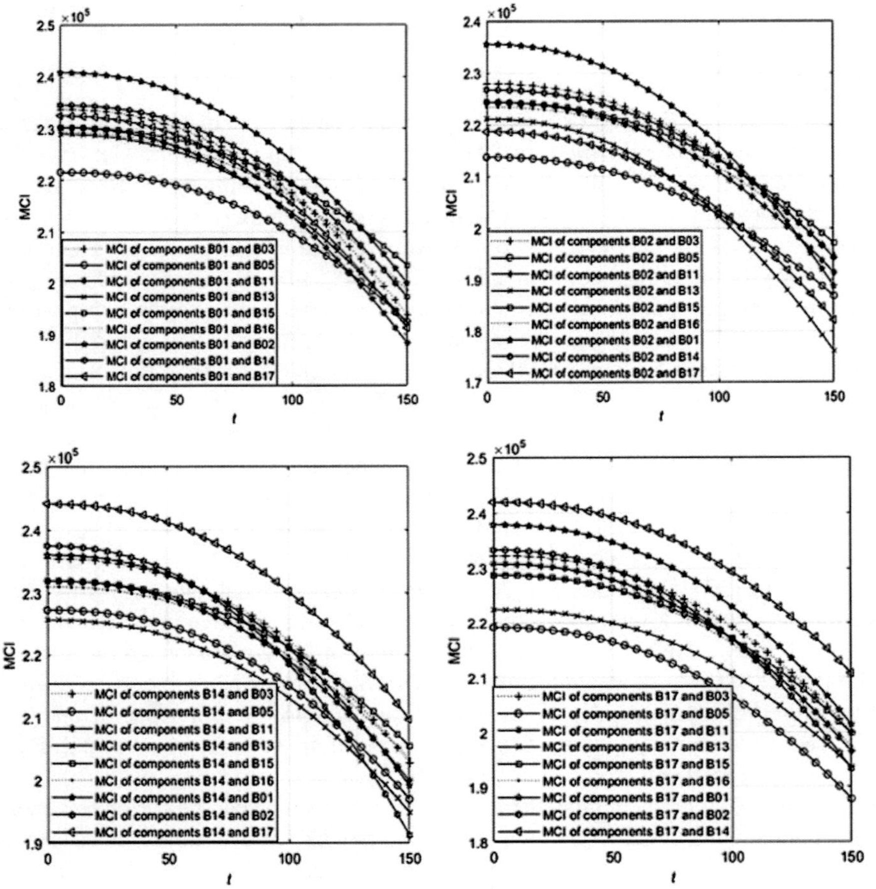

Fig. 7.25 CCMI when the critical component fails

on non-critical components that do not form a cut set with the failed component. For example, when non-critical component B03 fails, the components for PM at $t = 50$ are B05, B16, B13, B15, and B11, at which point the priority for PM for these components are B05, B16, B13, B15, and B11. and at $t = 90$, the priority for PM for the other components are B16, B05, B15, B13, and B11. If the non-critical component B15 fails, the components that can perform PM operations at $t = 50$ are B11, B13, B03, B05, and B16, at which time the priority for PM for these components are B11, B13, B03, B05, and B16. While at $t = 90$, the priority for PM for other components are B03, B11, B05, B13, and B16.

From the above analyses, it can be concluded that the priority of components ranked based on CCMI changes over time. The components selected for the PM may be different at different points in time. In addition, the components selected for PM

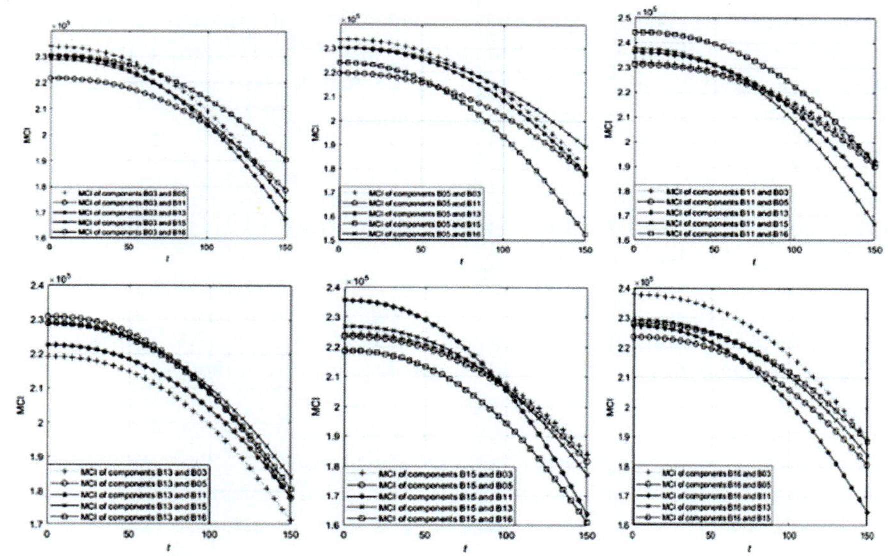

Fig. 7.26 CCMI when non-critical component fails

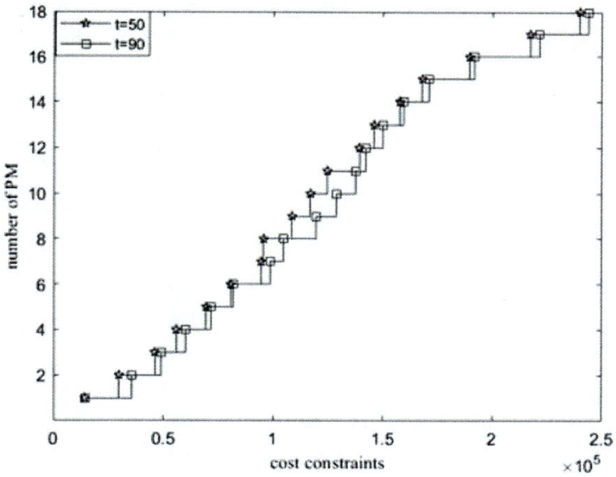

Fig. 7.27 Number of components for PM under the cost constraints

may be different if different components fail. This demonstrates the flexibility and usefulness of CCMI to provide repairmen with an optimal total repair cost for the priority of PM. Figure 7.27 shows the relationship between the number of components performing PM and the cost constraint when the critical component B01 fails.

As shown in Fig. 7.27, depending on the variation of the cost constraint, the number of components for PM that can be selected under this condition can be determined. For example, when $t = 50$ and the cost constraint is 100,000, the optimal number of components to perform PM is 8. When $t = 90$, the optimal number of components to perform PM is 7. This happens also due to the change in priority of components selected for PM. This shows that in practice, it is necessary to determine the number of components for PM at the right time due to the maintenance cost constraint, and provides some guidance for the maintenance policy of the RVATP.

7.4 Reliability and Repair Analysis of Complex Systems Under Multi-level Disasters

Engineered systems are nowadays becoming more complex. The dependencies between the components of a complex system are difficult to model, and complex systems are characterised by uncertainty [26]. The failure of a complex system can cause more damage to its environment and huge economic losses. As predicting the occurrence of failures is difficult and the damage caused by system failures is enormous, it is necessary to focus research on the reliability of complex systems and post-disaster analysis. This section examines post-disaster restoration strategies for complex systems under multi-level disasters. Based on the Markov model, the mechanism of the impact of component state changes on system performance is analysed, and the expected loss of complex systems is derived in consideration of the multi-level nature of failures. The post-disaster maintenance policies for complex systems is determined based on the estimated loss and the integrated importance measure.

7.4.1 Expected Loss Analysis of Complex Systems Under Multi-level Disasters Based on Markov Model

Suppose a system is composed of n components, each of which is numbered. The performance of component N_i is denoted by a_i. The system performance is a function of the component performances, represented by $Q = Q(a_i)$ with $i = 1, 2, \ldots, n$. It is assumed that all components work properly under normal conditions, and the system performance is degraded when a failure (or disaster) occurs. Denote $A' = \{N_i | 1 \leq i \leq n\}$ as the set of all components, $F' = \{N_i | 1 \leq i \leq f\}$ as the set of failed components, and $W' = \{N_i | f + 1 \leq i \leq n\}$ as the set of components in a working condition.

This section considers the multi-level nature of disasters. Suppose that the occurrences of the disasters are statistically independent and that the state transition process of the disaster level follows a homogeneous continuous time Markov process on the discrete state space. For different types of disasters, they are classified according to their nature.

Let the disaster level $\{\boldsymbol{X}(t), t \geq 0\}$ be a continuous-time Markov process in a discrete state space $\boldsymbol{E}'$ with $\boldsymbol{E}' = \left\{1, 2, \ldots, N'\right\}$, where 1 is the lowest level, N' is the highest level. The transition rate of a disaster level from i to j is u_{ij}, the transition rate matrix U of a disaster level $\begin{bmatrix} u_{11} & \cdots & u_{1N'} \\ \vdots & u_{ij} & \vdots \\ u_{N'1} & \cdots & u_{N'N'} \end{bmatrix}$. Let $P_j(T) = P\{X(T) = j\}$ with $j \in E$, which represents the probability that the disaster is at state j during time period $(0, T)$.

If a disaster u_k with level $X(t) = k(1 \leq k \leq N')$ occurs, the system performance of line N_i at time t becomes $a_i^k(t) = a_i$. The state of the line is defined by $h_i(t) \in \{0, 1\}$, indicating whether the line is failed or it is not at time t. If $N_i \in W', a_i^k(t) = a_i$, $h_i(t) = 1$, or $a_i^k(t) < a_i$, then $h_i(t) = 0$. The system flow at time t is denoted by $Q(k, t)$, it is a function of the flow of each component, as shown in Eq. (7.21).

$$Q(k, t) = Q\left[a_i^k(t)|N_i \in A\right]. \tag{7.21}$$

Each component in the system has M states, and the state transition of the component follows a Markov process.

Let $\{Y(t), t \geq 0\}$ be a stochastic process taking values on $E = \{1, 2, \ldots, M\}$. Then $\{Y(t), t \geq 0\}$ is said a continuous-time Markov process on the discrete state space E. The larger the value of $Y(t)$, the higher the performance. When M is taken, the performance reaches normal performance.

In the event of the k-th level disaster $(1 \leq k \leq N')$, let the state of the failed component N_i in the failed components set be denoted by $Y_i^k(t)$. The relationship between the state of the components and the actual performance is shown in Eq. (7.22), and ω is the state coefficient. Then

$$a_i^k(t) = \frac{Y_i^k(t)}{\omega} a_i. \tag{7.22}$$

The occurrences of a disaster and repairs of components cause the states of components to shift, and the transition rate matrix V of the components states is given by $\begin{bmatrix} v_{11} & \cdots & v_{1M} \\ \vdots & v_{ij} & \vdots \\ v_{M1} & \cdots & v_{MM} \end{bmatrix}$, where v_{ij} denotes the rate that the component departs from state i and ends in state j, $\sum_{j=1}^{M} v_{ij} = 0$, $i = 1, 2, \ldots, M$. When a disaster u_k occurs, it causes component N_i to transition from state M to state $Y_i^k(t)$, and the value of $Y_i^k(t)$ is any value in state space E. If the value of $Y_i^k(t)$ is m, the transition rate of this component from state M to state m is v_{Mm}, and the direct loss is d_{Mm}. Repair has improved the performance of failed components, transitioning from post disaster state $Y_i^k(t)$ to a better state. Suppose that a failed components can be repaired and transferred to the optimal state M and that only one component can be repaired at a time. Due to the need to limit the repair time to a time interval $(0, T)$, only some failed components will be selected for repair. The repair set for failed components is defined as R_i. Define a 0-1 variable R_i', with $R_i' = 1$ indicating that the failed component has been successfully repaired and N_i,$N_i \in R'$; A value of $R_i' = 0$ indicating that it will not be repaired, and $N_i \notin R'$.

The direct loss refers to the damage caused by the disaster to the infrastructure, which is only related to the post disaster state of the failed component. That is, it is the total cost required to restore the post disaster state $Y_i^k(t)$ to the optimal state M. It is assumed that the direct loss of failed component N_i under-level disasters is $L_i(k)$, the direct loss of the system under k-level disasters is the sum of the direct losses of all repaired components, as shown in Eq. (7.23). Due to the uncertainty of disaster levels, the expected direct losses of components and systems under different levels of disasters are used as the final direct losses of components and complex systems, as shown in Eqs. (7.24) and (7.25).

$$L_D(k) = \sum_{N_i \in F} (R'_i \times L_i(k)). \tag{7.23}$$

$$L_i = \sum_{k=1}^{N} L_i(k) \times P\{X(T) = k\}. \tag{7.24}$$

$$\begin{aligned} L_D &= \sum_{k=1}^{N} L_D(k) \times P\{X(T) = k\} \\ &= \sum_{k=1}^{N} \left[\sum_{N_i \in F} (R_i \times L_i(k)) \times P\{X(T) = k\} \right]. \end{aligned} \tag{7.25}$$

The indirect loss is caused by a system's performance degradation. Due to the uncertainty of the disaster level, assuming that the indirect loss of the system when a k-level disaster occurs is $L_{ID}(k)$, the expected value of the indirect loss at different levels is used as the final indirect loss L_{ID}, as shown in Eq. (7.26).

$$L_{ID} = \sum_{k=1}^{N} L_{ID}(k) \times P\{X(T) = k\}. \tag{7.26}$$

7.4.2 Repair Analysis of Complex Systems Under Multi-level Disasters

The system performance differs for different levels of disaster occurrence, as shown in Fig. 7.28. Taking the example of the kth level disaster occurrence, the changes in system performance can be explained as follows. The performance of a complex system during the normal operation is represented as Q_0. At time t_1, the kth level disaster occurs, causing the system performance to decrease to its lowest level. At time t_2, the repair of the failed components begins, and the repair process transitions the failed components from the post-disaster state to the optimal state, denoted by M.

Since only one failed component can be repaired at a time, the system performance gradually improves. At time t_3, the repair process is terminated, and the system performance recovers to $Q(k, t_3)$, which is equal to or lower than Q_0, where $Q_k < Q(k, t) < Q(k, t_3) \leq Q_0$. The shaded area represents the indirect losses $L_{ID}(k)$, of the system under the k-th level disaster. It is calculated as the integral of the difference between the normal system performance and the actual system performance post-disaster, as shown in Eq. (7.27). The total loss under the kth level disaster is denoted as Q_k^{loss} and represents the sum of direct losses and indirect losses, as shown in Eq. (7.28). The total loss of the complex system is represented by Q^{loss}, which is the sum of direct losses and indirect losses, as shown in Eq. (7.29). It can also be expressed as the expected value of losses, Q_k^{loss}, under different levels of disasters, as shown in Eq. (7.30).

$$L_{ID}(k) = \int_{t_2}^{t_3} (Q_0 - Q(k, t))dt. \tag{7.27}$$

$$\begin{aligned} Loss(k) &= L_D(k) + L_{ID}(k) \\ &= \sum_{N_i \in F} (R_i' \times L_i(k)) + \int_{t_2}^{t_3} (Q_0 - Q(k, t))dt. \end{aligned} \tag{7.28}$$

$$\begin{aligned} Q_k^{\text{loss}} &= L_D(k) + L_{ID}(k) \\ &= \sum_{N_i \in F} (R_i' \times L_i(k)) + \int_{t_2}^{t_3} (Q_0 - Q(k, t))dt. \end{aligned} \tag{7.29}$$

$$Q^{\text{loss}} = \sum_{k=1}^{N} \left[\left(\sum_{N_i \in F} (R_i' \times L_i(k)) + \int_{t_2}^{t_3} (Q_0 - Q(k, t))dt \right) \times P\{X(T) = k\} \right]. \tag{7.30}$$

During the repair process, the failed components can be repaired from the post-disaster state m to the optimal state M. The transition rate is denoted by v_{mM}, and the repair time for the components follows an exponential distribution with a rate of v_{mM}. The time required to restore from state m to state M is denoted by t_{mM}, which represents the expected repair time, calculated as $1/v_{mM}$. For the kth level disaster occurrence, the post-disaster state of the failed component N_i is denoted as $Y_i^k(t_1)$. The repair time for component N_i under the kth level disaster is denoted as T_i^k, as shown in Eq. (7.31). Considering the multi-level nature of disasters, the repair time for component N_i, denoted as T_i, is calculated as shown in Eq. (7.32). The total repair time for the entire complex system is denoted as T_{total}, corresponding to $t_3 - t_2$ in Fig. 7.28 as shown in Eq. (7.33). Additionally, the total repair time should be limited within the interval $(0, T)$.

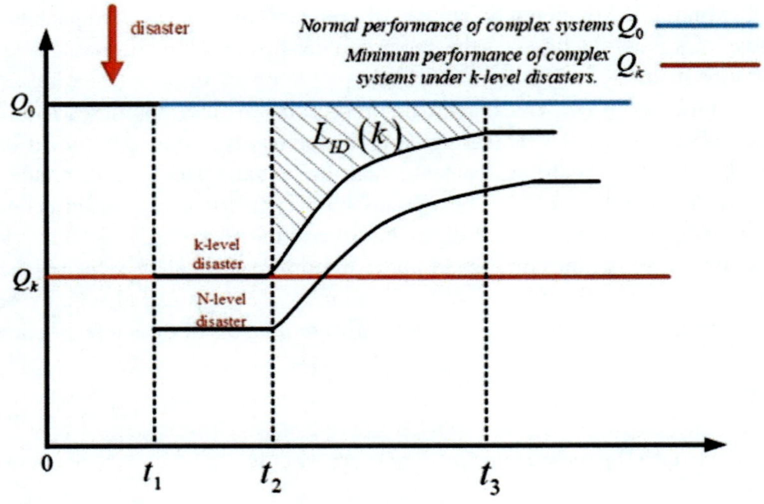

Fig. 7.28 System performance curves under different levels of disasters

$$T_i^k = 1/v_{Y_i^k(t_1),M}. \tag{7.31}$$

$$T_i = \sum_{k=1}^{N} T_i^k \times P\{X(T) = k\}. \tag{7.32}$$

$$\begin{aligned} T_{\text{total}} &= \sum_{N_i=1}^{f} [R_i'(T) \times T_i] \\ &= \sum_{N_i=1}^{f} \left[R_i'(T) \times \sum_{k=1}^{N} (1/v_{Y_i^k(t_1),M} \times P\{X(T) = k\})\right]. \end{aligned} \tag{7.33}$$

where $T_{\text{total}} \leq T$. The repair set of failed components is determined with the aim to minimise the expected total loss. Then the integrated importance measure of a complex system is defined, and the repair order of failed components is determined based on the integrated importance measure ranking.

$$\min \sum_{k=1}^{N} \left[\sum_{N_i \in F} (R_i' \times L_i(k)) \times P\{X(T) = k\} \right] + \sum_{k=1}^{N} (L_{ID}(k|R') \times P\{X(T) = k\}). \tag{7.34}$$

In Eq. (7.34), R' is the repair set of failed components, and $L_{ID}\left(k|R'\right) = \int_{t_2}^{t_3} \{Q_0 - Q\left(k, t|R'\right)\}\mathrm{d}t$ is the indirect loss of the system when the repair set is R' under the occurrence of the kth level disaster. $Q(k, t|R')$ denotes the performance of the system when the repair set is R'. Equation (7.34) indicates that determining the repair set of failed components can minimise the total loss.

The integrated importance measure of complex systems can be introduced, considering the transition intensity of component states, the multi-level nature of disasters, and losses. This concept aims to investigate the impact of component state changes on the overall system loss. The formulation is expressed in Eq. (7.35).

$$I_i^{\text{im}} = \sum_{k=1}^{N} P_{m_k}^i \times v_{m_k,\, M} \times P_k(T) \times \left[Loss_k(m_{k,i},\, R') - Loss_k(M_i,\, R')\right], \quad (7.35)$$

where m_k is the post-disaster state of component k when the kth level disaster occurs. $P_{m_k}^i$ denotes the probability that component N_i is in state m_k when the kth level disaster occurs. Since the post-disaster state of failed components is given in this section, $P_{m_k}^i$ is set to 1. $v_{m_k,M}$ is the transition rate of the component from the post-disaster state m_k to the normal state M. $P_k(T)$ denotes the probability of the kth level disaster occurring within the time interval $(0, T)$. $Q_k^{\text{loss}}(M_i, R)$ is the minimum total loss of the system when component N_i is repaired to the optimal state M under the kth level disaster. $Q_k^{\text{loss}}(m_{k,i}, R)$ is the minimum total loss of the system when component N_i is not repaired under the kth level disaster. I_i^{im} is the integrated importance measure of component N_i, which measures the extent to which repairing component N_i to the normal state reduces the total system loss. A higher value of I_i^{im} indicates a greater importance measure of the component, and its repair priority is higher.

7.4.3 IEEE18-Node Standard Power Distribution System Case Study

The IEEE 18-node standard power distribution system consists of 18 bus nodes with 17 lines. There are three types of nodes; the generation node is $N1$, the consumer nodes are $N10$, $N11$, $N14$, $N17$, $N18$, all of which have been labelled in Fig. 7.29, and the remaining nodes are distribution nodes. Each line represents a component and each grid branch consists of a number of lines, without considering the heat generated by the lines. The transmission capacity of each transmission line represents the component performance, the grid branch refers to as the route from the generation node to the distribution node and the sum of the maximum transmission capacity of each branch in the grid is taken as the performance of the power system.

There are three levels of a disaster, $P_j(t) = P\{X(t) = j\}$ $(j = 1, 2, 3)$ denote the probability of a line in state j. The transfer rate matrix U for the disaster level is given by $\begin{bmatrix} -0.075 & 0.075 & 0.0 \\ 0.025 & -0.05 & 0.025 \\ 0 & 0.075 & -0.075 \end{bmatrix}$. Suppose that the disaster level is constant over the time interval $(0, T)$ and the value of T is 16 days, the probability of a disaster staying at a level in $(0, T)$ is shown in Table 7.13.

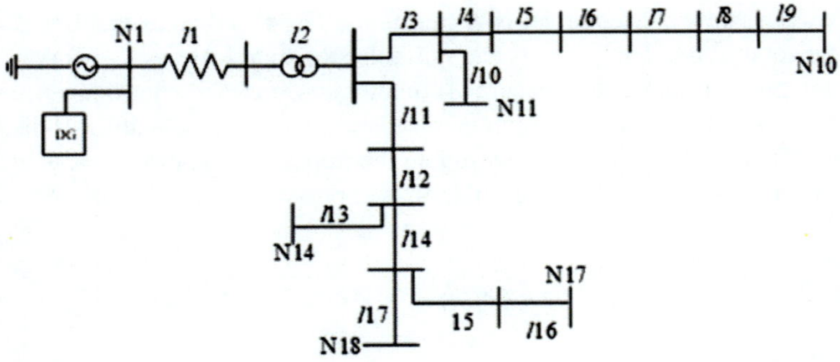

Fig. 7.29 IEEE 18-node standard distribution system

Table 7.13 Probability of a disaster being in a certain level in time period (0,T)

Disaster level	1	2	3
Probability	0.25	0.5	0.25

Assume that there are five states of the line, $Y_i^k(t) = \{1, 2, 3, 4, 5\}$, which suggests that the transmission capacity of the line becomes 1/5, 2/5, 3/5, 4/5, 1 of the normal transmission capacity respectively. The actual transmission capacity of the line can be derived from the line state by the relationship between the transmission capacity and the state $a_i^k(t) = \frac{Y_i^k(t)}{\gamma} a_i$, the line state transfer rate matrix is given by

$$\begin{bmatrix} -0.9 & 0.1 & 0.15 & 0.30 & 0.45 \\ 0.3 & -0.9 & 0.2 & 0.3 & 0.4 \\ 0 & 0.3 & -1.1 & 0.4 & 0.5 \\ 0 & 0 & 0.3 & -0.9 & 0.6 \\ 0 & 0 & 0 & 0.3 & -0.3 \end{bmatrix}.$$

Given the set of failed components at the time of a disaster by

$$F = \{l1, l3, l7, l8, l10, l12, l13, l14, l17\},$$

it is assumed that the set of failed lines at the time of the occurrence of a disaster with different levels is the same. The line states are shown in Table 7.14, where state 1 indicates the state of the line when a disaster of class 1 occurs, state 2 indicates the state of the line when a disaster of class 2 occurs, and state 3 indicates the state of the line when a disaster of class 3 occurs. The direct losses corresponding to the different hazards are shown in Table 7.15.

On the basis of the above data, we obtain the solution process shown below.

(1) Considering the repair time limit, the repair set R that satisfies the repair time limit is first found to have a total of 36 repair line sets as shown in Table 7.16.

Table 7.14 State of the failed lines

Line	$l1$	$l3$	$l7$	$l8$	$l10$	$l12$	$l13$	$l14$	$l17$
State1	1	1	4	4	3	3	2	2	4
State2	1	1	3	3	2	2	1	1	3
State3	1	1	2	1	1	1	1	1	2

Table 7.15 Direct loss of failed lines (RMB 10,000 Yuan)

Line	$l1$	$l3$	$l7$	$l8$	$l10$	$l12$	$l13$	$l14$	$l17$
L_i	2	4	3	3	1	4	3	3	2
L_i	2	4	4	5	1.5	6	5	4	3
L_i	2	4	5	6	2	7	5	4	4

Table 7.16 Repair line set and corresponding total loss (RMB, 10,000)

Repair line set R	Total loss	Repair line set R	Total loss
$R1: l7, l8, l10, l12, l13, l14, l17$	27.25	$R19: l1, l3, l8, l10, l12, l14, l17$	32.16
$R2: l3, l8, l10, l12, l13, l14, l17$	29.13	$R20: l1, l3, l8, l10, l12, l13, l17$	27.38
$R3: l3, l7, l10, l12, l13, l14, l17$	39.39	$R21: l1, l3, l8, l10, l12, l13, l14$	32.31
$R4: l3, l7, l8, l12, l13, l14, l17$	40.73	$R22: l1, l3, l7, l12, l13, l14, l17$	48.60
$R5: l3, l7, l8, l10, l13, l14, l17$	25.5	$R23: l1, l3, l7, l10, l13, l14, l17$	33.24
$R6: l3, l7, l8, l10, l12, l14, l17$	32.23	$R24: l1, l3, l7, l10, l12, l14, l17$	40.13
$R7: l3, l7, l8, l10, l12, l13, l17$	27.5	$R25: l1, l3, l7, l10, l12, l13, l17$	35.63
$R8: l3, l7, l8, l10, l12, l13, l14$	31.23	$R26: l1, l3, l7, l10, l12, l13, l14$	39.08
$R9: l1, l8, l10, l12, l13, l14, l17$	27.13	$R27: l1, l3, l7, l8, l13, l14, l17$	36.9
$R10: l1, l7, l10, l12, l13, l14, l17$	37.39	$R28: l1, l3, l7, l8, l12, l14, l17$	40.63
$R11: l1, l7, l8, l12, l13, l14, l17$	38.73	$R29: l1, l3, l7, l8, l12, l13, l17$	38.95
$R12: l1, l7, l8, l10, l13, l14, l17$	22.25	$R30: l1, l3, l7, l8, l12, l13, l14$	42.61
$R13: l1, l7, l8, l10, l12, l14, l17$	27.17	$R31: l1, l3, l7, l8, l10, l14, l17$	28.44
$R14: l1, l7, l8, l10, l12, l13, l17$	25.25	$R32: l1, l3, l7, l8, l10, l13, l17$	23.75
$R15: l1, l7, l8, l10, l12, l13, l14$	31.22	$R33: l1, l3, l7, l8, l10, l13, l14$	27.45
$R16: l1, l3, l10, l12, l13, l14, l17$	37.33	$R34: l1, l3, l7, l8, l10, l12, l17$	30.45
$R17: l1, l3, l8, l12, l13, l14, l17$	37.65	$R35: l1, l3, l7, l8, l10, l12, l14$	33.12
$R18: l1, l3, l8, l10, l13, l14, l17$	25.13	$R36: l1, l3, l7, l8, l10, l12, l13$	29.47

(2) The uniform direct and indirect losses are given in units of RMB 10,000 Yuan and let the unit price per kWh be RMB 1.0 Yuan. The total losses of the system under each repair set can be calculated separately, as shown in Fig. 7.30 and Table 7.16.
As can be seen from Fig. 7.30 and Table 7.16, compared to the other repair sets, when the repair set is $R12$, i.e., when emergency repairs are carried out on failed lines $l1, l7, l8, l10, l13, l14, l17$, the total loss to the system is the smallest, at

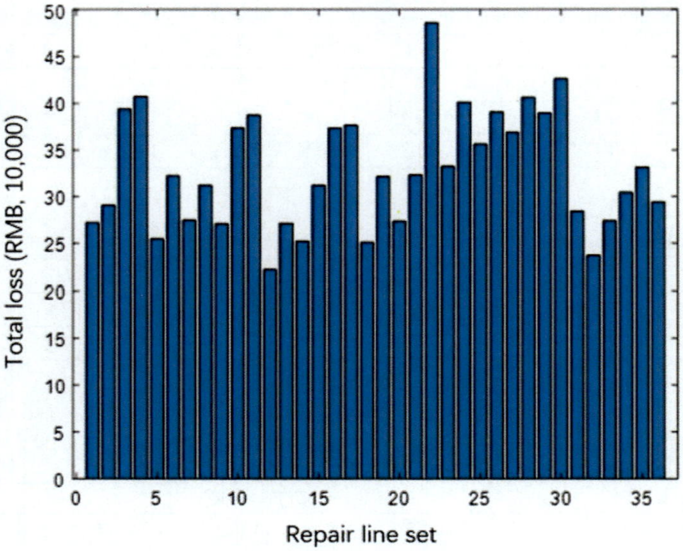

Fig. 7.30 Total loss under different repair line set

Table 7.17 The integrated importance measure values

Line1	Line7	Line8	Line10	Line13	Line14	Line17
0.9	2.799	8.7210	6.1226	4.7497	2.3250	4.1452

RMB 222,500 Yuan. Therefore, the failed lines $l1, l7, l8, l10, l13, l14$ and $l17$ are selected for emergency repair.

(3) Based on the integrated importance measure, the value of the integrated importance measure of each line can be found.
Due to the resource constraints, only one failed line can be repaired at a time, and different lines have different impacts on system losses, so the order of repair of failed lines $l1, l7, l8, l10, l13, l14$, and $l17$ must be determined. According to Eq. (7.35), the integrated importance measure value of a line is calculated and the importance measure value of the line can then be determined. According to the integrated importance measure model, the failed lines $l1, l7, l8, l10, l13, l14$, and $l17$ have integrated importance measures as shown in Fig. 7.31.

According to Fig. 7.31, the exact integrated importance measure values are shownn in Table 7.17.

Since the integrated importance measure characterises the impact of repairing the failed line on the total system losses, the greater the value of the integrated importance measure, the greater the reduction in total system losses due to repairing the failed line and the higher the repair priority of the failed line. Therefore, the repair priority of the failed line is $l8, l10, l13, l17, l7, l14$ and $l1$in descending order.

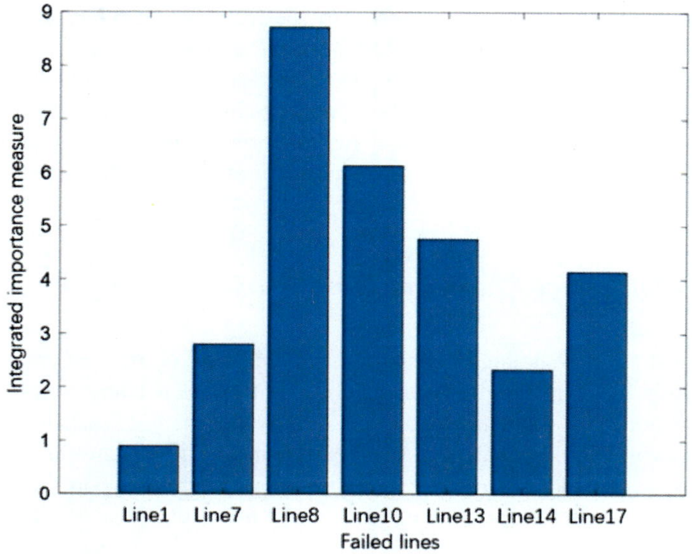

Fig. 7.31 Integrated importance measure of failed lines

7.5 Land Transport Network Systems

A complex network is a logical model composed of multiple nodes and edges. A land transport network is a typical complex network. Under the influence of human attacks, adverse weather, and natural disasters, when nodes and edges in the land transportation network fail, the integrity of the network will be damaged, and in severe scenarios, the entire network will be paralysed, which will incur unnecessary losses. Therefore, it is of great significance to study the reliability of land transportation networks and their ability to recover from interference and faults.

After the destruction of a land transportation network, in order to quickly recover network performance and improve network resilience, a key step is to identify the weak nodes in the network. Fang et al. [13] measured the criticality of network system nodes from the perspective of the contribution to system resilience. Cheng et al. [5] proposed a node importance assessment and network vulnerability analysis method based on a node state evolution cascading failure model. Cui et al. [6] proposed a resilience analysis model for network command information system based on node repair after the system is attacked. Feng et al. [14] proposed a key node identification method based on optimal network efficiency when a network system fails from an optimisation perspective. Huang et al. [19] proposed a comprehensive measurement method for node importance measures, considering the edge weight characteristics of a logistics network. Yan et al. [27] studied the vulnerability and measurement indicators of a complex networks. Jalan et al. [20] used network centrality to identify key nodes in an ocean transportation network. Chen et al. [3] proposed an improved method to identify key stations that affect the performance of urban rail networks. Deng et al. [7]; Sun et al. [24] introduced the Tabu search algorithm to optimise the network robustness measure to identify key nodes for the optimal attack Policy.

This section considers the loss and recovery of land transportation network performance with importance, and proposes loss importance, recovery importance, and resilience importance models to measure the importance of transportation nodes.

7.5.1 Performance Change of Land Transport Network

The transport node in the land transport network is the stop of the transport activities, which is the starting point and the end point of the network transport route. A transport node is a standstill place to undertake the material distribution, warehousing, transit, and deployment of the work. Transport routes are the sides of the land transport network, and transport activities on different transport routes need to be connected through transport nodes. If transport nodes or routes are damaged, the transport capacity of the land transport network will also be reduced. Let us define the performance $A(t)$ of a land transport network as the ratio of the freight volume of all non-failed nodes in the land transport network after being attacked at time t to the freight volume of all normal nodes in the land transport network before the attack. $A(t) \in [0, 1]$. $A(t)$ varies with the actual freight volume of nodes in the land transportation network, and can be used to measure the transportation capacity of the land transportation network. During the recovery of the failed node, the performance of the land transport network gradually recovers.

In order to maximise the transportation performance of the entire land transportation network during operation, the network flow is distributed according to the principle of the maximum freight volume of the entire network during the entire process, and the following model is established.

$$\max A(t) = \max \frac{\sum_{i \in N_u} q_i(t)}{\sum_{i \in N} q_i(t)} \tag{7.36}$$

$$q_i(t) \le \tau_i(t) C_i, \quad i \in N \tag{7.37}$$

$$q_{ij}(t) \le \tau_{ij}(t) C_{ij}, \quad ij \in L \tag{7.38}$$

$$\tau_{ij}(t), \tau_i(t) \in \{0, 1\}, \ (i, j) \in N, t \in \forall t, \tag{7.39}$$

where N is the set of all transport nodes in the land transport network, N_u is the set of non-failed transport nodes in the land transport network, $q_i(t)$ is the freight volume of transport node i at time t, $q_i(t) \le C_i$ is that the freight volume of the node is less than or equal to the freight capacity of the node, $q_{ij}(t)$ is the freight volume on the transport route ij at time t, $q_{ij}(t) \le C_{ij}$ means that the freight volume of the transport route is less than or equal to the capacity of the transport route, $\tau_{ij}(t)$ and $\tau_i(t)$ are the states of the transport route ij and the transport node i at time t, respectively, $\tau_{ij}(t) = 1/\tau_i(t) = 1$ indicates that transport route ij or transport node i is in normal condition, and $\tau_{ij}(t) = 0/\tau_i(t) = 0$ indicates that transport route ij or transport node i is in a failed state. Further, with the resilience importance presented in Sect. 6.3, we can obtain

$$\begin{aligned} I_i^{\text{resilience}}(t) &= \frac{I_i^{\text{recovery}}(t)}{I_i^{\text{loss}}(t)} \\ &= \frac{Q^{\text{recovery}}(t)_{\tau_i=1} - Q^{\text{recovery}}(t)_{\tau_i=0}}{Q^{\text{recovery}}(t)_{\tau_i=0} - Q^{\text{recovery}}(t)_{\tau_i=1}}, \end{aligned} \tag{7.40}$$

where $I_i^{\text{resilience}}(t)$ refers to the resilience importance measure of each transport node in the land transport network, which is used to evaluate the impact of different nodes on the network performance. $\max\{I_i^{\text{resilience}}(t),\ i = 1,\ 2, .., n\}$ is the transport node which has the greatest impact on the performance of the land transport network.

7.5.2 Case Study

In this section, the nodes in a land transport network represent the transport distribution nodes and transport dispersion nodes. The edges represent the transport routes between the nodes. The transport nodes include six transport distribution nodes and 11 transport distribution nodes. The six transport distribution nodes include Qingdao City, Ningbo City, Shanghai City, Tianjin City, Guangzhou City, and Chongqing City in China. The eleven transport dispersion nodes are Changsha City, Wuhan City, Zhengzhou City, Nanjing City, Hangzhou City, Nanchang City, Beijing City, Xi'an City, Jinan City, Chengdu City, and Hefei City in China. There are 28 transport routes in total. The logical route of the land transport network is shown in Fig. 7.32. The parameters of each transport node and freight volume at the initial time are shown in Table 7.18.

In Fig. 7.32, the network flow is represented by the amount of goods of nodes and edges. The state of all transport nodes and lines in the land transport network is independent of each other. The transport distribution nodes NS = {S1, S2, S3, S4, S5, S6}. The transport dispersion nodes ND = {D7, D8, D9, D10, D11, D12, D13, D14, D15, D16, D17}. The transport routes L = {S1D9, S1D15, S2D11, S2D12, S2D17, S3D10, S3D11, S3D12, S4D9, S4D13, S5D7, S5D12, S6D7, S6D8, S6D14, S6D16, D7D8, D7D12, D8D12, D8D17, D9D14, D9D15, D9D16, D9D17, D10D17, D11D17, D13D14, D14D16}.

Suppose that the transport nodes {S1, S2, S4, D7, D9, D11, D14, D17} fail. According to Eq. (7.36), only if the transport node D7 fails, the network performance $A(t_b)_{\tau_7=0} = 0.9391$. According to Eq. (7.36), only if transport node D7 fails, the loss performance value is $\text{loss}(t)_{\tau_7=0} = A_{\text{target}}(t) - A(t_b)_{\tau_i=0} = 1 - 0.9391 = 0.0609$. Similarly, only if the transport nodes S1, S2, S4, D9, D11, D14 and D17 fail, the performance values are 0.9542, 0.9439, 0.9456, 0.9544, 0.9502, 0.9578, 0.9438,

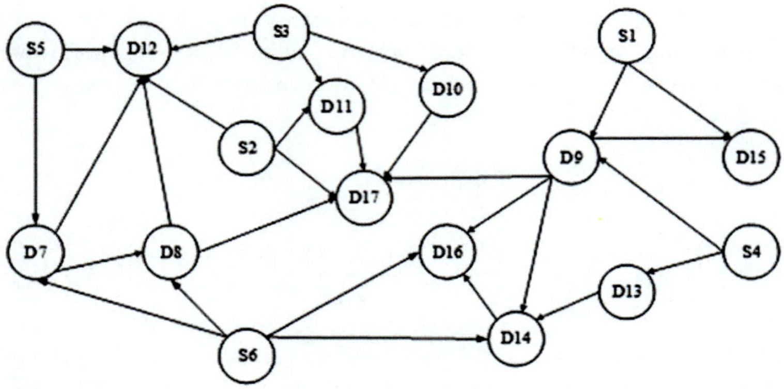

Fig. 7.32 The logical route of the land transport network

Table 7.18 Transport node and freight volume at the initial time

Serial number	Transport node	Freight volume $q_i(t_0)$
1	Qingdao city	29170
2	Ningbo city	35757
3	Shanghai city	38750
4	Tianjin city	34711
5	Guangzhou city	88351
6	Chongqing city	89965
7	Changsha city	38808
8	Wuhan city	41778
9	Zhengzhou city	29039
10	Nanjing city	16886
11	Hangzhou city	31732
12	Nanchang city	15350
13	Beijing city	22325
14	Xi'an city	26901
15	Jinan city	34000
16	Chengdu city	28207
17	Hefei city	35858

respectively. The loss of the performance values of the land transport network when only the failed transport nodes are not working properly are obtained respectively, shown in Table 7.19.

When all transport nodes are working properly, the land transport network is in its initial state. Therefore, the loss performance value is $\mathrm{loss}(t)_{T_i=1} = 0$. The loss importance values of the failed transport node are shown in Table 7.20.

Table 7.19 Loss performance when the state of the transport node is 0

Transport node	S1	S2	S4	D7	D9	D11
Loss performance	0.9542	0.9439	0.9456	0.0609	0.9544	0.9502
Transport node	D14	D17				
Loss performance	0.9578	0.9438				

Table 7.20 The loss importance value of the transport node

Transport node	S1	S2	S4	D7	D9	D11
The loss importance	0.0458	0.0561	0.0544	0.0609	0.0456	0.0498
Transport node	D14	D17				
The loss importance	0.0422	0.0562				

In Table 7.20, the greater the loss importance value of the transport node, the greater the impact of the failure of the transport node on the performance recovery of the land transport network. Therefore, it can be concluded that the influence degree of nodes on the vulnerability of land transport network is D7, D17, S4, S2, D11, S1, D9 and D14 in turn.

Each failed node in the land transport network can be recovered within the same recovery time. However, only one failed node can be restored within a recovery period. The transport nodes and freight volume after recovery are shown in Table 7.21.

According to Eq. (7.36), when the failed nodes are not restored to its normal state, the lowest network performance $A(t_b)$ is 0.5408. For the failure node, only if node D7 is restored to its normal state, the network performance is $A(t_\gamma)_{T_7=1} = 0.6003$. According to Eq. (7.36), only the recovery performance of the failure node D7 after it is restored to its normal state is $\text{recovery}(t)_{T_i=1} = A(t_\gamma)_{T_7=1} - A(t_b) = 0.6003 - 0.5408 = 0.0595$. Similarly, the performance values of transport nodes S1, S2, S4, D9, D11, D14 and D17 are 0.5887, 0.5957, 0.5941, 0.5855, 0.5734, 0.5823, 0.6709, respectively. The recovery performance of the land transport network only if the failed transport node i returns to normal operation are obtained, as shown in Table 7.22.

When all nodes are not back to their normal operating states, the performance of the land transport network is in the worst state. Therefore, when the transport nodes are not back to normal, the recovery performance value of the land transport network is $\text{recovery}(t)_{T_i=0} = 0$. According to Eq. (7.40), the recovery importance measure of the failed transport node is shown in Table 7.23.

In Table 7.23, the greater the recovery importance of the transport node, the greater the impact of node failures on the performance loss of the land transport network. Therefore, it can be concluded that the degrees of the influence of nodes on the elasticity of land transport network are D17, D7, S2, S4, S1, D9, D14 and D11 in turn.

Table 7.21 Transport nodes and freight volume after recovery

Serial number	Transport nodes	Freight volume $q_i(t_\lambda)$
1	Qingdao city	31120
2	Ningbo city	35757
3	Shanghai city	59750
4	Tianjin city	34711
5	Guangzhou city	67351
6	Chongqing city	77860
7	Changsha city	38808
8	Wuhan city	53883
9	Zhengzhou city	29039
10	Nanjing city	16536
11	Hangzhou city	21032
12	Nanchang city	10350
13	Beijing city	22325
14	Xi'an city	26901
15	Jinan city	31070
16	Chengdu city	18601
17	Hefei city	85458

Table 7.22 The recovery performance when the transport node state is 1

Transport node	S1	S2	S4	D7	D9	D11
Recovery performance	0.0479	0.0549	0.0533	0.0595	0.0447	0.0326
Transport node	D14	D17				
Recovery performance	0.0415	0.1301				

Table 7.23 The recovery importance measure of the transport node

Transport node	S1	S2	S4	D7	D9	D11
The recovery importance	0.0479	0.0549	0.0533	0.0595	0.0447	0.0326
Transport node	D14	D17				
The recovery importance	0.0415	0.1301				

According to the node loss importance measure and recovery importance measure the resilience importance measure of the failed node is obtained, as shown in Table 7.24.

In Table 7.24, the higher the resilience importance measure of the nodes, the greater the impact of node recovery on the resilience recovery of the land transport network, and the higher the node recovery priority. Therefore, the priority order of transport node recovery is D17, S1, D14, D9, S4, S2, D7 and D11.

Table 7.24 Resilience importance measure of the transport node

Transport node	S1	S2	S4	D7	D9	D11
Resilience importance	1.0852	0.9786	0.9798	0.9770	0.9803	0.6546
Transport node	D14	D17				
Resilience importance	0.9834	2.3149				

7.6 Summary

This chapter introduced some applications of importance measures in hydraulic tension systems, submarine blowout preventer systems, aircraft Hydraulic systems, unmanned aerial vehicles, industrial robot systems, wind power generation systems, electricity power grids, the main coolant system of nuclear reactors, highway and maritime transport networks, metro networks, healthcare supply chain networks, and equipment support networks, respectively.

References

1. Birnbaum ZW (1968) On the importance of different components in a multicomponent system. Washington Univ Seattle Lab of Statistical Research, Tech. rep
2. Cavalcante CA, Lopes RS, Scarf PA (2018) A general inspection and opportunistic replacement policy for one-component systems of variable quality. Eur J Oper Res 266(3):911–919
3. Chen T, Ma J, Zhu Z, Guo X (2023) Evaluation method for node importance of urban rail network considering traffic characteristics. Sustainability 15(4):3582
4. Chen YL, Chang CC, Zhang ZG, Chen X (2020) Optimal preventive 'maintenance-first or-last' policies with generalized imperfect maintenance models. J Ind Manage Optim 17(1):501–516
5. Cheng G, Lu Y, Zhang M, Huang J (2017) Node importance evaluation and network vulnerability analysis on complex network. J Nat Univ Defense Technol 39(1):120–127
6. Cui Q, Li J, Wang H, Nan M (2018) Resilience analysis model od networked command information system based on node repairability. Comput Sci 45:117–121
7. Deng Y, Wu J, Tan Y (2016) Optimal attack strategy of complex networks based on tabu search. Phys A: Stat Mech Appl 442:74–81
8. Dui H, Si S, Yam RC (2017) A cost-based integrated importance measure of system components for preventive maintenance. Reliab Eng Syst Saf 168:98–104
9. Dui H, Wu S, Zhao J (2021) Some extensions of the component maintenance priority. Reliab Eng Syst Saf 214:107729
10. Dui H, Zheng X, Wu S (2021) Resilience analysis of maritime transportation systems based on importance measures. Reliab Eng Syst Saf 209:107461
11. Dui H, Yang X, Fang Y (2022) Evaluation methodology for preventive maintenance in multi-state manufacturing systems considering different costs. Int J Prod Res 1–16
12. Dui H, Zheng X, Guo J, Xiao H (2022) Importance measure-based resilience analysis of a wind power generation system. Proceed Inst Mech Eng Part O: J Risk Reliab 236(3):395–405
13. Fang Y, Pedroni N, Zio E (2016) Resilience-based component importance measures for critical infrastructure network systems. IEEE Trans Reliab 65(2):502–512
14. Feng X, Hu C, Xu C et al (2019) Key node reconition method based on optimal network efficiency. Comput Eng Design 40(02):328–335

15. Finkelstein M, Cha JH, Bedford T (2023) Optimal preventive maintenance strategy for populations of systems that generate outputs. Reliab Eng Syst Saf 237:109334
16. Gao K, Peng R, Qu L, Wu S (2020) Jointly optimizing lot sizing and maintenance policy for a production system with two failure modes. Reliab Eng Syst Saf 202:106996
17. Geng J, Li Y, Wang R, Wang Z, Lv C, Zhou D (2017) A virtual maintenance-based approach for satellite assembling and troubleshooting assessment. Acta Astronaut 138:434–453
18. Hashemi M, Tavangar M, Asadi M (2022) Optimal preventive maintenance for coherent systems whose failure occurs due to aging or external shocks. Comput Ind Eng 163:107829
19. Huang Y, Jin C, Rong L (2014) Cascading failure model on logistics network based on the overall importance of nodes. Oper Res Manage Sci 23:108–115
20. Jalan S, Yadav A, Sarkar C, Boccaletti S (2017) Unveiling the multi-fractal structure of complex networks. Chaos Solitons Fract 97:11–14
21. Nasrfard F, Mohammadi M, Karimi M (2023) A petri net model for optimization of inspection and preventive maintenance rates. Elect Power Syst Res 216:109003
22. Perrons RK, Richards MG (2013) Applying maintenance strategies from the space and satellite sector to the upstream oil and gas industry: a research agenda. Ener Policy 61:60–64
23. Shi Y, Lu Z, Huang H, Liu Y, Li Y, Zio E, Zhou Y (2022) A new preventive maintenance strategy optimization model considering lifecycle safety. Reliab Eng Syst Saf 221:108325
24. Sun Y, Yao P, Zhang J, Fu K (2017) Node attack strategy of complex networks based on optimization theory. J Elect Inf Technol 39(3):518–524
25. Wei S, Nourelfath M, Nahas N (2023) Analysis of a production line subject to degradation and preventive maintenance. Reliab Eng Syst Saf 230:108906
26. Xiaoping L, Yiming T, Liping Z (2008) Survey of complex system and complex system simulation. J Syst Simul 20(23):6303–6315
27. Yan K, Li L, Qin Y et al (2018) Key node identification methods based on road traffic networks. Comput Eng Sci 40(11):1983–1990
28. Zhao X, He Z, Wu Y, Qiu Q (2022) Joint optimization of condition-based performance control and maintenance policies for mission-critical systems. Reliab Eng Syst Saf 226:108655
29. Zheng R, Zhou Y, Zhang Y (2020) Optimal preventive maintenance for wind turbines considering the effects of wind speed. Wind Ener 23(11):1987–2003
30. Zhu X, Chen Z, Borgonovo E (2021) Remaining-useful-lifetime and system-remaining-profit based importance measures for decisions on preventive maintenance. Reliab Eng Syst Saf 216:107951